A LAYMAN'S GUIDE TO
SPACE, TIME, THE UNIVERSE
AND OTHER UNBELIEVABLE STUFF

NICK SPINDLER

I AM SELF-
PUBLISHING

@iamselfpub
http://www.iamselfpublishing.com

To Lindy Lou for all her patience.

CONTENTS

"We are all in the gutter, but some of us are looking at the stars."

(Oscar Wilde, *Lady Windermere's Fan*)

PREFACE

This is the book I've always looked for on the science bookshelves but have never been able to find. So I decided, somewhat boldly, to write it myself.

Got a science background? I hear you say. Nope. Ever written a book? That's also a no. Ever written anything? Not that I remember. And there you have it – the perfect credentials.

Because this book is for all us folk who are amazed, dumbfounded and 'mouth-open gaping' at all the unbelievable space and universe stuff we hear and read about but are, somehow, unable to grasp. The proverbial penny won't drop. And that's because the people who write the vast majority of popular science books are scientists and physicists. Brilliant, yes, but they are so far up the knowledge ladder they've forgotten what it's like to be scrabbling around just trying to reach the first rung. Even the books that profess to be 'for the beginner' start with an assumption of basic knowledge that the average man and woman in the street just doesn't possess.

That's what this book aims to address. To explain, as simply as it is possible to do, concepts such as space and time, the nature of black holes, the birth and death of stars and the size of the visible universe. Things of gigantic proportions, and things so small they have no proportions. The impossible, unimaginable and unbelievable, explained by a layman. So I, for one, will finally hear that penny go 'clink' and, hopefully, so will you.

Most of the facts, figures and wonderful analogies I refer to are down to the miracle that is Google and the various shades of scientific genius Google so effortlessly serves up. (I have credited you all and I thank you.)

What I've done is to try to organise stuff in a half-logical manner, pick out the bits that I'm most fascinated by (or in some instances, least understand) and then wrap a little of my own language around it, in an attempt to make it simple.

Doing 'simple' is bound to leave a few gaps for those of you who know better. So to you, all I can ask is you take heart that even if only a few people read this book that hadn't previously thought too much about space, time and the universe, then a few more people will have moved a little closer to the stuff you love so much.

And that can't be a bad thing.

So, here we go, to boldly go where no layman has gone before.

Wish me luck! (And you, of course.)

BIG NUMBERS

Where to start? Well, if we're going to be able to get a feel for the scale of the visible universe and the monstrous size of the things that inhabit it, then we're going to have to get a feel for big numbers. And this involves getting our heads around numbers like billions, trillions, quadrillions and other equally hard to remember 'illions'.

Take an octillion, for example; it looks like this (take a breath!): 1,000,000,000,000,000,000,000,000,000.

Isn't it amazing how just a simple nought can suddenly become so mind-boggling when there's 27 of them lined up after a '1'? And are you ever likely to encounter an octillion of anything? Well, if you're an average-sized person, you're a walking, talking, living example of what seven octillion atoms look like when they get together. So, yes. But we'll talk atoms in Chapter 4; let's deal with the names and numbers of noughts first.

1 million is two sets of three noughts –
1,000,000

1 billion is three sets of three noughts –
1,000,000,000

1 trillion is four sets of three noughts –
1,000,000,000,000

1 quadrillion is five sets of three noughts –
1,000,000,000,000,000

1 quintillion is six sets of three noughts –
1,000,000,000,000,000,000

I won't go on. As you can see, I've expressed each number in a way I find it easier to remember, which is the number of sets of three noughts it has. As you will also see, the sets of three noughts are always one more set of three than the name suggests. You think of a trillion (tri) as meaning three (e.g. tricycle, triangle), but it has four sets of three noughts. You would think of a quadrillion (quad) as meaning four (e.g. quad bike), but it has five sets of three noughts. And so on. A set of three noughts more than the name suggests.

If someone said to me now, 'How many noughts in a quintillion?', my tiny brain would go 'quint' is five (e.g. quintet, quintuplets), which means six sets of three noughts. (Six times three sets of noughts = 18. Hurrah!)

Is anyone ever likely to ask? Probably not. But it feels nice to be on top of stuff like this, to be able to get it and remember it, even if it's just for your own personal satisfaction. (How many noughts in an octillion? See how fast you can get there.)

PS – It's on p.11, if you want to check it out.

We can also express these figures in language, not noughts:

1 million is a thousand thousand

1 billion is a thousand million

1 trillion is a thousand billion

1 quadrillion is a thousand trillion

1 quintillion is a thousand quadrillion

Simple enough – times the last big number by a thousand and you get to the next big number. But just how big are these? What does a billion look like, feel like? Or a trillion? Well, there are quite a few analogies out there that try to help us grasp the gigantic scale of these numbers. Here are a few that grab me:

1 second	=	**1 second** (simple enough)
1 million seconds	=	**12 days** (not too concerning)

1 billion seconds	=	30 years (where did that come from! That is a frightening jump)
1 trillion seconds	=	30,000 years (one trillion seconds ago the pyramids hadn't been built!)
1 quadrillion seconds	=	30,000,000 years (yes that is indeed 30 million years!)
1 quintillion seconds	=	30,000,000,000 years (oh dear, that's 30 billion years, over twice the length of time the universe has been in existence!)

If ever an analogy exposed the unbelievable size of these numbers as they get bigger and bigger, this is it. The difference between a billion seconds and a trillion seconds being in the region of 30,000 years is just mind-blowing.[1]

Another one. There are, as you may know, many billionaires in the world. (Lucky? Criminal? Rich parents? Who knows? But they have it.) If one of these billionaires took it upon themselves on a rainy Sunday afternoon to count their billion pounds (or dollars or any other currency), it would take them around 265 years to go from one pound to one billion pounds (based on a 16-hour counting day, with time off for food, bed and the toilet, of course). How big is a billion? It's 265 years' worth of your life, or a billionaire's life, to be more exact. Just to count your money. (Please don't try this at home.)

That's a billion. But what if you attempted to count just to the next big number – a trillion? It doesn't look that far away in the list... one billion... one trillion... right next to each other.

The answer? In the region of 542,241 years! (That's five hundred and forty-two thousand, two hundred and forty-one years, just to be clear.) And you thought 265 years to count to a billion was bonkers. How is this number worked out?

It's based on each lower set of numbers taking an average of three seconds to pronounce, with up to 19 seconds average as you get to the really, and I mean *really*, big numbers, e.g. 963,736,324,632. That's nine hundred and sixty-three billion, seven hundred and thirty-six million, three hundred and twenty-four thousand – what's on telly? – six hundred and thirty-two.[2] How many numbers would you have to read out in total? 369, 472, 888, 227. (You know by now this is billions.)

Staying with money, take a five-pound note, a dollar bill, or any other paper currency. Each of these is about 0.0043 inches thick (0.10922 mm). This is a useful analogy because we can envisage a paper note. We can even get one out of our purse or wallet and look at it while reading this. For analogies, familiar things are good. Now you know where we're going.

1 x five-pound note
sits
0.0043 inches 'high' (0.10922 mm). Tiny.

1 thousand x five-pound notes
sit
4.3 inches high (10.922 cm). I can separate my thumb and middle finger more than this.

1 million x five-pound notes
sit
358 feet high (109.118 metres). We've certainly gone past body part analogies now – think as high as St Paul's Cathedral, if you know it.

1 billion x five pound notes
sit
68 <u>miles</u> high (109.435 km). That's miles! We're now close to the edge of the Earth's atmosphere and on the verge of space! And there are billionaires out there. Where do they keep it all?

1 trillion x five-pound notes
sit
<u>68 thousand miles high</u> (109,435.39 km). This is what a trillion does to a tiny, innocuous five-pound note just 0.0043 inches high. It takes it into space and a quarter of the way to the Moon.

1 quadrillion x five-pound notes

sit

<u>68 million miles high</u> (109,435,392 km). We've just created a 'paper pile' that's shot past Mars by thirty-odd million miles.

1 quintillion x five-pound notes

sit

<u>68 billion miles high</u> (109,435,392,000 km). Forget planets, this particular stack of 'fivers' has left the solar system! And continued on and upwards for another 64 billion miles or so. And they're just £5 notes, 0.0043 inches high!

If piles of money don't quite do it for you, see www.ehd.org – Grasping Large Numbers. They not only do money-pile analogies, but 'money carpets' and 'money lines', each one trying to help us grasp these ridiculously big numbers.

So, that's our warm-up exercises on just how big, big numbers are. You might now be able to grasp the scale of world debt, which currently sits at a knee-trembling 160 trillion dollars – a ten-million-eight-hundred-and-eighty eight-thousand-mile-high stack of dollar debt!

In the following chapters, we will be using distances of billions and trillions of miles, and be describing things as, for example, 'a quadrillion times bigger', in our attempt to grasp distances and scales that are the stuff of madness. Now and again you might come back here, just to remind yourself of the vast leaps in 'bigness' that happen between these numbers, so the numbers don't just roll over you in some vague state of 'illions' fatigue. A billion seconds is 30 years. A trillion seconds is 30,000 years. A quadrillion seconds is 30 million years. A quintillion seconds is 30 billion years – more than twice the age of the universe!

Now, we only went to a quintillion, but these big numbers do just get bigger and bigger and... BIGGER.

Take the 'googol' (sound familiar?). If a quintillion has 18 noughts behind the '1', the googol has this many:

10,000,000,000,000,000,000,000,000,000,000,000,000,00
0,000,000,000,000,000,000,000,000,000,000,000,000,000
,000,000,000,000,000,000,000!

Count them all? That's 100 noughts. (The exclamation mark is my little contribution.)

The amazing thing about this number is it has no physical reality. There isn't a googol of things anywhere in the universe. There is no one object you could count that would ever add up to the enormity of this number – not grains of sand on Earth, not stars in the universe, not even, quite astonishingly, all the atoms in the visible universe. In fact, if you filled the whole of the visible universe with individual grains of sand, all packed tightly together to fill every cubic millimetre of space, you still wouldn't have a googol of sand grains. You would have to multiply all that 'universe worth' of sand grains 10 billion times (!) to get close to a googol of sand grains.[3]

And who came up with the name 'googol'?

A nine-year-old boy by the name of Milton Sirotta. He was the nephew of a Dr Kasner, who, when writing a book entitled *Mathematics and the Imagination* (Kasner and Newman, 1940), asked the young Milton to make up a name to best describe a very big number. And the rest is history; other than a minor and totally genuine spelling mistake made by some bloke called Larry Page – googol – Google?

Does anything lie beyond the vastness of the googol? You bet! A centillion. That's a '1' followed by 330 noughts. (I'll let you write this one out.) How big is this? Well, it's 1,000 googols, of course. (Come on, keep up!)

Does having a number like this make any sense? Of course not, but you have to marvel at the sheer size of it. And, would you believe, the young Milton wasn't done with just the googol. Flushed with success (or possibly vexed that a centillion dwarfed his googol) he proposed an even bigger number – the 'googolplex' – a '1' followed by a 'googol of noughts'. (Imagine that many noughts in a nine-year-old's head!)

Finally (almost), bonkers maths people created an even bigger number... the googolplexian (what the...?). That's a '1' followed by 'a googolplex of noughts'! They are actually fun numbers to look at, if only to go boss-eyed scrolling down the endless rows of noughts. Go and do it at www.

googolplex.com and www.googolplexion.com. (See you in a couple of thousand years.)

I said 'almost' because there is a number that's even bigger than a googolplexian. It's called 'Graham's Number' and was invented in 1977 by a still-living American scientist called Ronald Graham. His number is SO BIG that the really, really clever people, including Ronald himself, can't even tell you how many noughts it has!

Now, that's what you call BIG.

The table overleaf contains a list of big numbers, from a million up to a googolplex. The little number by the '10' says how many noughts follow the '1', e.g. a quintillion is 10^{18}. Each big number is multiplied by a thousand to make the next big number, which in turn is multiplied by a thousand, and so on, to create just the most wonderfully, impossibly big numbers.

If anybody ever asks you, 'What's a googol?' put your hand on your chin, rummage around a little, looking very circumspect, and then answer, 'Hmmm… I think that will be around ten duotrigintillion.' (Ha ha!!)

So, with this in our pocket, let's go travel the universe!

BIG NUMBERS TO IMPOSSIBLY BIG NUMBERS[4]

NUMBER	NAME	NUMBER	NAME
10^6	Million	10^{42}	Tredecillion
10^9	Billion	10^{45}	Quattuordecillion
10^{12}	Trillion	10^{48}	Quindecillion
10^{15}	Quadrillion	10^{51}	Sexdecillion
10^{18}	Quintillion	10^{54}	Septendecillion
10^{21}	Sextillion	10^{57}	Octodecillion
10^{24}	Septillion	10^{60}	Novemdecillion
10^{27}	Octillion		And onwards...
10^{30}	Nonillion	10^{99}	Duotrigintillion
10^{33}	Decillion	10^{100}	Googol
10^{36}	Undecillion	10^{303}	Centillion
10^{39}	Duodecillion	$10^{(10^{100})}$	Googolplex

CHAPTER 2

HOW FAR IS FAR?

That's big numbers. Now, let's start to use some of them to try to get a grip on just how big the visible universe is.

The best way I've found to get my layman's head around the enormity of the distances involved is to think about how long it would take us to travel those distances at the fastest speed a manned spacecraft has ever gone through space. And be warned, now, if you have any kind of inferiority complex, it's going to grow big time as you come to realise just how puny and totally insignificant we truly are. So, how fast can 'we' currently travel through space? Well, the space speed record was set quite some time ago, on 26 May 1969, to be exact.

On that day, Eugene Cernan, Tom Stafford and John Young, the crew of Apollo 10, were rocketing back to Earth from the Moon (as you do) and reached the quite amazing speed of 24,791 mph (39,897 kph).

At that speed, Apollo 10 was covering seven miles, through space, every second (rounded up from 6.8863 miles a second). Try to imagine somewhere roughly seven miles from where you are now. Now imagine getting there in basically a blink of your eye. (Blink that eye again and try to feel that kind of speed.) On that May day in 1969, that's exactly what the crew of Apollo 10 were doing. A speed of seven miles a second, which is truly astounding.

How does this rank against our Earthly experiences, so we have something to compare it to?

The fastest movement most of us will ever experience is in a commercial airliner, which flies somewhere around 500–600 mph (804–965 kph), depending on conditions. If we allow for the top end of 600 mph, that's just under one-fifth of a mile every second, versus seven miles a second for the crew of Apollo 10. Imagine your jumbo suddenly accelerating to 41 times its top speed. (Fasten seat belts!)

Before they were decommissioned, Concorde, and a Russian-based equivalent called the Tupolev Tu-144, had maximum cruising speeds of 1,354 mph (2,179 kph) and 1,510 mph (2,430 kph) respectively. That's roughly three times as fast as you in your jumbo, but still only half a mile a second, versus seven miles a second (so not even close). At these speeds, both Concorde and the Tupolev were travelling at twice the speed of sound (Mach 2.02). The fastest speed recorded to date by a manned aircraft (versus spacecraft) is 4,473 mph (7,198 kph). This was achieved by a very brave man (to my mind), William J. 'Pete' Knight, in the X-15 rocket jet on 3 October 1967. 'Pete' went around seven times faster than your 'fatty jumbo' and travelled at one and a quarter miles a second, which is more than six times the speed of sound (Mach 6.70).

When the crew of Apollo 10 did their 24,791 mph they were not just covering seven miles per second (versus Pete's one and a quarter miles) they were travelling at an astonishing 32 times faster than the speed of sound! (When you next talk to someone, say a friend, and you experience what feels like their instantaneous reaction to your words, think about Apollo 10 – it was travelling 32 times faster than your voice to your friend's ears. I mean, wow!!)

Forty-nine years later, the crew of Apollo 10 still holds the record for the fastest a manned spacecraft has ever gone in space. So, to get a feel for cosmic distances, let's see just how far this current 'manned space speed record' would get us, were we able to replicate that speed consistently. (It's worth noting here that the 24,791-mph speed record was only achieved for a certain period of time on re-entry into Earth's atmosphere. It's not a consistent average speed. Rockets, of course, have to take off, get up to speed, etc. But for the purposes of this exercise, we will use this record speed as if we were able to consistently produce it, just to keep things simple.)

THE MOON

Let's set our sights on something reasonably familiar, so we can get our bearings regarding distance and our ability to cover it.

We've been to this little fella, the Moon, a few times. The first was on July 20–21 1969, when Neil Armstrong and Buzz Aldrin first landed. That was followed by five other manned US landings.

And for those of us who haven't made the journey – quite a number, I'm guessing – we get a pretty good look at the Moon reasonably regularly. We can stop on occasions, stare at it across the dark void of space and at least get a feel for how far away it is.

The Moons distance from Earth varies. At its closest (known as 'perigee') it's about 225,622 miles away (363,103 km) and when it's at its most distant from Earth (known as 'apogee') it's about 252,088 miles away (405,696 km).

These distances reflect the fact that the Moon orbits the Earth in an elliptical pattern, which means at certain points it will be closer to us and at other points further away.

For the sake of argument, let's assume we can fly through space at a consistent speed of that achieved by Apollo 10, over the shortest possible distance to the Moon, so we give ourselves every advantage we can. This is what you get:

The shortest distance to the Moon	–	225,622 miles
Divided by the fastest man has ever travelled in space	–	24,791 miles per hour
If we fly at this speed, consistently and straight, it will take us	–	About 9 hours

In human terms, nine hours is nothing, even allowing for the nine hours back. So, space travel is looking pretty good at this familiar distance.

The actual fastest trip, allowing for all the realities of take-off, trajectory, landing and so on, was completed by the very famous crew of Apollo 11, Neil Armstrong and Buzz Aldrin among them. They went from Earth into a lunar orbit in around 2 days, 3 hours and 49 minutes, returning in another 2 days, 22 hours and 56 minutes. That's a round trip travelling through space of just over 5 days. As you can see by this real example of travelling to the Moon, I've given us an enormous 'unreal' advantage by assuming we can travel at a consistent speed of 24,791 mph with an 18-hour round trip, versus the reality of a 5-day round trip. But, as you will also see, this advantage will count for nothing as we go further and further.

TO MARS

So far, so good. Let's move onto a little planet that is still reasonably familiar, in as much it's the subject of numerous science fiction books and films. And we have actually landed space probes there. The first successful fly-by of Mars was way back on the 14–15 July 1965, by NASA's Mariner 4. On 14 November 1971, Mariner 9 became the first space probe to enter into the Mars orbit.

The first to make contact with the surface of Mars were two Soviet probes – Mars 2 landed on 27 November 1971 and Mars 3 landed on 2 December 1971. In the ensuing years many other missions have followed, with two rovers currently on Mars and beaming back signals to Earth – 'Spirit' and 'Opportunity' of the Mars Exploration Rover (MER) mission and a larger rover, 'Curiosity' of the Mars Science Laboratory (MSL) mission. And, as of 2018, there are six active craft currently orbiting and surveying planet Mars.

And a man or woman on Mars? Both NASA and other bodies, such as the European Space Agency, are making active plans with a view to launching the first manned mission to Mars sometime between 2030 and 2035. A word of warning, however; don't go rushing to volunteer for this little excursion. Roughly two-thirds of all unmanned spacecraft destined for Mars over the last fifty years have failed before completing their mission (two-thirds!). Clearly, space travel is a tricky business, even beyond the issue of distance.

Anyway, undaunted to Mars we go!

Just as the Moon follows an elliptical path around the Earth, both Earth and Mars follow elliptical patterns around the Sun. So, once again, we get the potentially shortest distance to Mars (called 'perihelion') – a mere 33.9 million miles (54.5 million km). However, the two planets haven't been this close in recorded history. The closest recorded so far happened in 2003, at 34.7 million miles (55.8 million km), so we will use this 'real' figure. Out of interest, the furthest the Earth and Mars can be apart is when they are in 'opposition'; that is, on opposite sides of the Sun (called 'aphelion'). The average distance between the two planets at this point is 249 million miles (400 million km)! As you can see, when sending someone to Mars, you want to be doing it when Mars is close.

So, to 'close' Mars we go:

The shortest distance to Mars from Earth	–	34.7 million miles
Divided by the fastest man has ever travelled in space	–	24,791 miles per hour
If we fly at this speed, consistently and straight, it will take us	–	58 days

Based on our fastest 'space speed', without worrying about the complicated realities of accelerating, slowing down, the type of trajectory, etc., we have a handy 58-day journey. Pretty good for space travel to another planet. Factor in the realities when we do launch a manned spacecraft to Mars and it's reckoned the actual journey time will be around 260 days. Again, using our fastest-speed figure, as if we could do it consistently, really does give us quite an advantage. Trust me, we're going to need it.

BEYOND, TO PLUTO

So, we got to the Moon in around 9 hours and to Mars in around 58 days. Let's take a bigger step, avoiding all the other planets in our solar system, and point the nose of our super-fast spacecraft to Pluto, the furthest planet from Earth; right at the edge of our solar system. Two quick points to note here. There are many differing opinions on what defines the edge of our solar system. For the sake of this exercise I'm defining it as just beyond Pluto. There is also, and will probably continue to be, some debate about whether Pluto really stacks up as a true planet. Officially, it's been decreed it no longer does. None of this, though, matters much for our purposes; we're still heading there as it's the furthest round thing we can land on before we shoot off into interstellar space!

Just as Earth and Mars follow an elliptical pattern around the Sun, so does Pluto. Once again, we have a short and a long measure of distance. At its most distant, when Earth and Pluto are at opposite sides of the Sun, the distance between here and there is a whopping 4.2 billion miles (6.7 billion km). We've moved from millions of miles to a number that represents a thousand million (a billion) and, in this case, there are four of them.

Thankfully, like before, we will take the shortest distance and that's a much friendlier 2.66 billion miles (4.28 billion km).

So, start the engines:

The shortest distance to the furthest 'planet' in the solar system, Pluto	–	2.66 billion miles
Divided by the fastest man has ever travelled in space	–	24,791 miles per hour
If we fly at this speed, consistently and straight, it will take us	–	12 years (or thereabouts)

That's starting to get to a quite considerable amount of time, bearing in mind we are also giving ourselves a considerable consistent-speed advantage. However, 12 years in a lifetime? It's not too bad a time to travel to the furthest 'planet' in our solar system (not forgetting it's 12 years back, too).

And we've been to Pluto already, or rather a probe has. NASA's 'New Horizons' was launched on 19 January 2006 and reached (and then flew by) Pluto on 14 July 2015. That's 9 years, 5 months and 25 days, because 'New Horizons' is unmanned. As a consequence, it reached speeds during its journey of around 52,000 mph (83,000 kph) – twice as fast as man has ever travelled.

And, if you're wondering what guesstimates of time are being made to get to Pluto on more realistic average spacecraft speeds, you get – shortest distance to Pluto, 2.66 billion miles, divided by the average speed a manned spacecraft can currently do, 17,500 mph, which gives you around 17 years there and 17 years back. (Half a lifetime. Imagine being brave enough to set off on a 34-year round trip.) Again, you see the advantage of using the fastest speed ever achieved as a consistent speed.

So, we're done with our solar system, simply because there are no other planets beyond Pluto to visit. Now, I'm afraid, we are going to have to leave it. I say 'afraid' because what lies beyond the comfortable confines of our own solar system is a nerve-shredding chasm of unimaginable size and volume. It's the place where the very big numbers live. The trillions, quadrillions and quintillions, and on to the kind of numbers that will drown your senses.

Let's go there anyway, if perhaps a little 'less boldly'.

The next nearest star is the obvious target, for the simple reason, there's nothing nearer. It's called Proxima Centauri and is part of what is called the 'Alpha Centauri triple-star system'. And, amazingly fortunately for us, in this system scientists have identified a little planet orbiting the star Proxima Centauri, called Proxima b. Now, this is where all thoughts of familiarity disappear, like envisaging the distance to the Moon as it hangs low in the sky on a beautiful summer's night. Or feeling 'at one' with the red planet, because Matt Damon's been there.

Because the star Proxima Centauri and its little orbiting planet, Proxima b, (the next one on from Pluto), are 25 trillion miles away from Earth (40 trillion km), we've lost millions of miles and quickly dispensed with billions of miles. We're now talking trillions. We have 25 trillion miles (that's 25 thousand, billion miles) to navigate our way across, just to get to the next nearest star and nearest known planet outside our solar system. At this distance we will forget elliptical patterns and the like and just deal with the figure in hand. It's enough to make your legs go hollow.

Hold tight...

The distance to the nearest planet in the nearest star system – Proxima b	–	25 trillion miles
Divided by the fastest man has ever travelled in space	–	24,791 miles per hour
If we fly at this speed, consistently and straight, it will take us (gulp!)	–	Around 115,000 years!!!!

Yes. Let me write that out so the penny drops. Even at the fastest man has ever gone in space, the journey to just the nearest star and nearest planet outside our solar system will take one hundred and fifteen thousand human years. That's close on 4,000 generations, just to cross 'the gap' from the edge of our solar system to the next nearest star and planet.

If you were to arrive there now, in 2018, you would have to have set off just as the first *Homo sapiens* were moving out of North Africa and about to bump into Neanderthal man. That's how long it would take you. Because that's how far it is just to the nearest star and planet beyond Pluto.

There may be a 'wall' for you here. There certainly was for me. When I first saw this figure – one hundred and fifteen thousand years – just to travel to the nearest star and planet outside of our solar system, I found myself mulling it over and over, trying to grasp how we could jump from a potential 12-year trip to Pluto, which was feeling so good, to 115,000 years as our next stop! We're travelling at seven miles per second, for heaven's sake, nearly 25,000mph!

But, I suppose, that's the purpose of this exercise and of this little book. To find a way to place ourselves within the ridiculously gargantuan thing we call the universe. Most of us, I think it would be fair to say, live the gift we've been given, myopic to the reality around us. In this respect, we are not unlike Jim Carrey's character, Truman Burbank, in the film, *The Truman Show*.

Poor old Truman goes from birth to adulthood totally not knowing that the blue sky above his head is just a giant façade – part of a TV reality show film set. How many of us are any different? We live our lives scurrying around, busy-busy, never contemplating what lies just outside our little blue bubble world. Truman Burbank eventually awakens to his reality and takes a journey across tempestuous seas to confront and go beyond the blue facade, with the immortal words to the millions watching on TV, 'Thank you and goodnight.' Just like Truman we should, on occasion, let our minds bang on the blue facade above our heads, and go beyond it to 'experience' what's out there.

Let your mind cross the 225,000-mile, 9-hour journey to the Moon. Stretch the 34-million- mile, 58-day journey to the red planet. Leap the 2.66-billion-mile, 12-year voyage to little Pluto. And then, standing on the very edge of the solar system, with Pluto at your back, look out to that faintest of glows in deep, deep space, Proxima Centauri and its little planet, Proxima b, and, in the deafening silence, stare into the maw of the abyss that separates you from that tiny glow. An airless void, so unimaginably vast it would take 115,000 human years to cross. Just to get to the little pinprick of light that's the next nearest star and planet. And, if standing on the edge of that abyss isn't enough to fill you with awe or terror, look even further into the dark, beyond Proxima Centauri, and you will begin to see an even more distant glow, then another and another, until you start to become conscious of a vast ocean of stars swamping your vision, as our galaxy, the Milky Way, reveals itself in all its glory... the light of 400 billion stars. And just you. On your tiptoes. On the edge.

Breathtaking.

The actual journey time to Proxima Centauri, were we ever in a position to attempt it, would be more like 150,000 years. It really is mind-boggling. And what's the next nearest star, beyond Proxima Centauri, out of curiosity? It's called Barnard's Star and it lies around 35 trillion miles from Earth – an extra 10 trillion miles on from the Alpha Centauri triple-star system and its little planet, which adds a small matter of roughly another 60,000 to 70,000 years of travel time.

A crumb of comfort to ease us past the chasm waiting just outside our solar system: we can see the Alpha Centauri star system and Barnard's Star in the night sky with the naked eye, just like the Moon, Mars and Pluto. A wafer-thin vestige of familiarity we can hold onto.

So, where to from here? (Home, maybe?) Well, it probably isn't a good idea to keep stretching out into space, going star to star, planet to planet, for the simple reason, as we've already noted, there are around 400 billion stars in our galaxy, the Milky Way, and it's estimated there are around 500 million planets of one sort or another scattered amongst them. Going one at a time would make for a mighty long book for my first attempt. So, we're going to jump from the nearest star and planet to the galaxy itself as our next measure of distance.

And such is the scale of this jump that we're going to have to change our measure of distance. If we stay in miles, we will reach numbers with so many zeros your eyes will boggle, and you will probably start dribbling. So, like most folk do at this stage, we will switch to a measure called 'light years'. A little explanation, before we move on.

To help describe the vast distances we are now going to try to get our heads around, we will refer to the distance light can travel in a year, to create a measure of distance called a 'light year'. Just how far can light travel in a year? Unbelievably far. Light is the fastest thing we know of in the universe. It travels at:

186,282	miles per second	(that's thousands)
11,176,920	miles per minute	(millions)
670,615,000	miles an hour	(millions)
16,094,764,000	miles a day	(billions)
5,900,000,000,000	miles a year	(trillions)

In just one year, light can travel 5.9 trillion miles! To destroy any sense we had of being able to go pretty fast ourselves, thank you, let's compare what we can currently do at our fastest space speed. 'We' can travel at:

7	miles per second	(that's 7)
413	miles a minute	(hundreds)
24,791	miles an hour	(thousands)
594,984	miles a day	(thousands)
209,434,368	miles a year	(millions)

You can make the comparisons yourself, but let me just point one out. In one year (what we now refer to as a light year) light can travel 5.9 trillion miles (remember just how big a trillion is). To cover that same distance at our maximum speed of 24,791 mph would take us in the region of... 28,000 years! This, I hope, gives you some indication of the speed of light. So, when we talk from now on in light years, keep in mind we are talking about each light year being close to six trillion miles in distance, something that would take us 28,000 years to cover! (It's worth coming back to this now and again as we move on to the unfathomable scale of the reality around us.)

For a first easy reference point for using light years as a measure of distance, let's just go back to Proxima Centauri. The distance to Proxima Centauri of 25 trillion miles, is just 4.3 light years. As you can see, it's a lot easier than saying 'twenty-five trillion miles'. But it can make distances look small if you don't constantly keep in mind what one light year means in actual distance.

ACROSS THE MILKY WAY

So, across the galaxy we go, undeterred. Let's have the size of it straight out, in light years, and then try to deal with it. Our galaxy, the Milky Way, with its 400 billion stars, is one hundred thousand light years across. That's from one side to the other – 100,000 light years!

Stop and think for just a moment. We've seen that light can cover the most astonishing distance in just one year – 5.9

trillion miles. But the Milky Way galaxy, which we are part of, is so humongously vast that light itself takes a mind-blowing one hundred thousand years to make one crossing! And that's only there, not there and back.

For a specific stream of light to just be crossing the end of our galaxy now, it would have to have set off at a time when there were as few as 10,000 people on planet Earth, versus the current 7.6 billion. What chance do we have of ever crossing just our own galaxy? (Other than absolutely none, so end the book here and now!)

Well, however ridiculous, let's try to figure it out and see what we get.

The distance across the Milky Way	–	100,000 light years
Divided by the fastest man has ever travelled in space (oh dear)	–	24,791 miles per hour (get a move on!)
If we fly at this speed, consistently and straight, it will take us	–	2,800,000,000 years (yes, that's billions!)

Just how big is the galaxy we are in? It takes light, which can travel at the most ludicrous speed of 5.9 trillion miles a year, 100,000 years to get across our galaxy once. For us, wanting to make the same journey, going at the fastest speed man has ever gone in space, it's a mind-bending 2.8 billion years! That is over half the time our planet, Earth, has been in existence, and roughly a quarter of the time the universe has been in existence. And it's only our galaxy!

This book is about trying to get some kind of perspective on our place in the universe. And if 2.8 billion years, just to cross the distance of our own galaxy doesn't do it, then not much will (and we're using a 'cheating speed'!). We think we're technologically advanced, clever, a superior species. And yes, compared to other species on the planet we've not done badly. But relative to the galaxy we live in, the Milky Way, we're a shrivelled midge, beyond minuscule. Not even embryonic in terms of scale or significance. So, what do we have so far, in terms of distance to cover and the time it will take us to cover it?

To the Moon	–	9 hours (get me a ticket!)
To Mars	–	58 days (take a sabbatical)
To Pluto	–	12 years (mmm... and back?)
To the nearest star and planet	–	115,000 years (anyone want a cheap ticket?)
To go across our own galaxy	–	2.8 billion years (anyone?)

And still we need to go on. Just as we asked what the nearest star system and planet is to our solar system, so we should also ask, 'What's the nearest galaxy?' Because we are 'boldly layman', after all. It's called the Andromeda Galaxy, and would you believe, it's twice as big as our galaxy, the Milky Way, coming in at a whopping 220,000 light years across. In terms of distance from our galaxy, it's a mind-curdling 2.5 million light years. By now, you will have a feel for a light year distance, but just to emphasise the point – a light year is 5.9 trillion miles, so the Andromeda Galaxy is 5.9 trillion x 2.5 million miles away. Let's do the journey thing for the last time. (I can hear you in the back going, 'Are we there yet?')

The distance to the nearest galaxy, the Andromeda Galaxy, from our galaxy, the Milky Way	–	2.5 million light years
Divided by the fastest speed man has ever gone in space (d'oh!)	–	24,791 miles per hour
If we fly at this speed, consistently and straight, it will take us	–	Around 70 billion years (no exclamation marks, it's just too ridiculous)

When you travel in a jumbo jet you will reach speeds of around 600 mph and think you're going pretty fast. It puts into perspective the achievement of the Apollo 10 crew, and the people who put them up there, to be able to achieve a speed of close on 25,000 miles per hour – seven miles every second. But then you think of the nearest galaxy to ours, and that speed is diminished to a point beyond insignificance when you calculate it would take around 70 billion years to make the journey. Close to five times longer than the universe itself has been in existence! That's how far, far is. And that's just to the galaxy next door.

Will we ever get to visit our local neighbour? Well, help is at hand, because the Andromeda Galaxy, in a friendly and neighbourly fashion, is coming to say hello to us. And quite enthusiastically, it would appear, as it's hurtling towards the galaxy we live in, the Milky Way, at an incredible speed of 244,800 miles per hour (that's 68 miles/109 km per second!).

That's some going for a galaxy 220,000 light years across and containing around one trillion stars. What's driving this propulsion? The attractive power of gravity that the two galaxies exert on each other (more on gravity in Chapter 5). Will it hit us? Seemingly, most definitely. And it's only now you begin to give thanks for the 2.5 million light years that separate us. It's a distance beyond comprehension in terms of our own ability to travel it, and, thankfully, a challenging one even for Andromeda.

Even at a speed of 244,800 miles an hour it will take Andromeda five billion years to reach us, versus the 70 billion years for us to travel to it. (Think of the timescales involved here. Think of the distances involved. And we are only talking about our galaxy and the next nearest galaxy, both tiny blobs of nothing in the scheme of things.) Out of interest, what happens in five billion years' time when the Andromeda Galaxy, with its one trillion stars, hits our Milky Way, with circa 400 billion stars? Well, this is likely to astound you, but not a lot. The chances of any two stars actually colliding are negligible! However unbelievable that sounds, it's evidence again of the huge distances each galaxy covers and, as a consequence, the huge distances between individual stars.

On average, the distance between individual stars is around 100 billion miles (160 billion km). A good analogy – imagine putting down a ping-pong ball, walking two miles and putting down another ping-pong ball. That's a good representation of the average distance between stars, making individual collisions when the galaxies merge highly unlikely. You have to shake your head at the sheer 'awesomeness' of it all.

Finally, what happens to us? Well, according to current predictions by two scientists at the Harvard–Smithsonian Center for Astrophysics, there's a 50 per cent chance our solar system will be swept out three times further from the galactic core than we are at present and a 12 per cent chance we could be 'ejected' from the new galaxy altogether![5] (That will be the second gulp! of the book.) The good news? Even if we were ejected, the event would likely have no adverse effect on the Sun, planets or the balance of our solar system (phew!).

The bad news? By the time the two galaxies collide, the surface of the Earth will already have become far too hot for liquid water to exist, thus ending all life on Earth (oh…). Make a note in your diary, that's in about 4.75 billion years' time.[6][7]

So, after all that, what do we have? We're unlikely to ever visit the Andromeda Galaxy, with its journey time of 70 billion years. And, for all its sterling efforts, the Andromeda Galaxy's 244,880 mph race to visit us is all to no avail, as life on Earth ceases to exist just as Andromeda is about to knock on the door and say 'hello'.

The next time you are out at night, lying in the gutter with a clear night sky above your head, look up and take in all the stars you can see. They look familiar because we're used to seeing them, but we're not used to thinking about them. The light from just the nearest star outside our solar system has taken four and a half years to reach your eyes. That's how long it takes light, travelling at 186,282 miles a second, to cross the chasm that separates you and Proxima Centauri. You are, in fact, looking at Proxima Centauri as it was four and a half years ago. You are looking back in time. And if that isn't amazing enough, the little star you are looking at is an unfathomable 25 trillion miles away. A distance so vast that even travelling at the fastest speed man has ever gone in space, it would still take in excess of 115,000 years to cross. And that's just the nearest star.

Look a little harder and you will be able to see the light of what is thought to be the most distant star visible with the naked eye, a star called V762 Cas, in the constellation of Cassiopeia, some 16,000 light years away (around 94 quadrillion miles). The light you are looking at from V762 Cas started out just as our cave-dwelling ancestors were getting to grips with the idea of making clay pots. It travelled through and out the other end of the Ice Age, saw the creation of the Sahara Desert, watched the Roman Empire, and all of the empires that followed, rise and fall. It saw the painting of the Sistine Chapel, witnessed Hitler come close to taking over the world, and heard the voice of Martin Luther King. And it was still travelling as you were born.

Then, one night, as you raise your face to the sky, that long-travelling little light beam plops into your eye. And in that moment a distance of 16,000 light years opens up in front of you, and you see the star V762 Cas, in Cassiopeia, as it was 16,000 years ago. Truly unbelievable.

And if you wanted to make the journey in the opposite direction, at the fastest speed man has ever gone, to the furthest star you can see with your naked eye?

Close to 450 million years.

That's how far, far is.

CHAPTER 3

HOW BIG IS BIG?

So, we got to Andromeda… or, to be more exact, Andromeda got to us. And within all of this we've begun to get a feel, or a glimpse, of the gargantuan monster that is the thing we call the universe. And yet, it's only the distance to Andromeda, which on the scale of the universe is the tiniest of steps (a mere 2.5 million light years). So again, let's not bother with any intermediate steps, like the two trillion other galaxies that are thought to inhabit the visible universe (that's two thousand billion!). Let's just jump instead to the visible universe itself.

Now, the best advice when trying to comprehend the size of the visible universe is, best not. You can't. It's not actually possible. It's so unimaginably vast that our brains can't compute the distance and volume involved. Why? Because in all of our human history, all two million or so years of it, we've gained no experience of such things. We walk miles.

We can envisage a mile, two, ten even. Many people know what it's like to run 26 miles. Then there are motoring miles. We can sense what a 200-mile journey will be like. Guess the time pretty accurately. We can grasp air miles; we have a pretty good feel for London to New York or vice versa, at 3,500 miles, around seven hours.

But that's about it. We have no reference point to judge, say, a hundred thousand miles, half a million or a million miles. Nothing we have ever learned or experienced gives us an innate sense of what these distances feel like. The only people who have got anywhere near close to distances like this are those astronauts who have travelled the 225,000 miles to the Moon and the 225,000 miles back. And they are few. So, to understand the size of the visible universe, we are going to have to try to imagine the unimaginable. Something we are not wired to do.

How do we go about it?

Well, there is one analogy I came across that half-works for me, written by a Bradford G. Schleifer, at www.rcg.org.[8] I say 'half-works' because we are talking about the unimaginable. This is beyond anything your brain will ever have to comprehend. And, for all the trying, like me, you'll probably never quite land it. But, on brief occasions, you might just get a tiny glimpse. You'll know when that happens, as you'll be suddenly gasping for air. It goes like this, with an apology, because it's going to feel a bit slow and laboured. Big things normally are.

Imagine Earth. Picture it as we see it in photographs from space. That huge, hanging blue sphere, close to 8,000 miles in diameter (that's across) and 24,000 miles in circumference (that's around) and weighing one thousand trillion metric tonnes (that's just bloody heavy). Now, in your head, reduce it to a grain of salt. A single, incy wincy grain. To make it more real, if you can, go to your kitchen and drop a single grain of salt into the palm of your hand. See our planet in all its humongousness and then reduce it down to that grain of salt, with you in there somewhere. It's a big reduction, a 42.5 billion times reduction, to be exact. We are now going to describe the scale of the universe on this new 42.5 billion reduced scale, with the grain of salt representing 'us' – planet Earth.

Here we go.

Step 1

With your little grain of salt in the palm of your hand (or imagined in your head) go into a reasonably sized room in your house or flat. Sit or stand in the middle of the room and look around you. 'Feel' the volume. Now look down at the grain of salt in your hand. If this room were only the visible universe and the salt grain was Earth, we already appear shudderingly tiny (Don't just read this, go do it).

Step 2

Now step outside. Or do this when you're going to work, college, shopping. Look up at the sky as it arcs above your head. The great volume of blue, stretching up 62 miles to the edge of space. When you feel this volume around you, above, left, right… look down to the palm of your hand to the barely visible, by comparison, grain of salt. Imagine the outer edges of that blue sky not just above your head, but circling around and down, 62 miles below you (as the universe is a sphere).

Take all of this huge volume in, and then look at 'us' – the little salt grain. It should take your breath away. And then imagine this 124-mile-diameter sphere (62 miles up and down) is thick black. That should make your knees tremble.

And it's just a sphere, 124 miles in diameter. If this were only the scale of the universe in comparison to Earth, it would be big enough and scary enough for me. Try to hold onto this scale, because it's about the only reference point that will give you some idea of what small, puny and insignificant really means. Where we're going, our planet will remain at this scale – a grain of salt. But the sphere around it is going to grow and then grow some more, until it becomes impossible to hold the mind-numbing reality of it in your head.

Step 3

With the tiny grain of salt (Earth) nestled in your shaking little hand, now try to imagine this 124-mile-diameter sphere starting to expand all around you, doubling to 248 miles in diameter, then again to close on 500 miles, again to over 1,000 miles, again and breathless again until it rests at 8,000 miles in diameter! The little salt grain ('us') is now suspended in the middle of a huge black sphere the size of Earth itself at the full Earth-size.

Can you imagine we could be so small? A tiny 0.3 mm grain of salt in comparison to a visible universe the size of the whole of planet Earth. If only this were it.

Step 4

Now imagine this Earth-size sphere expanding to encompass the Moon. Again, something you may be able to imagine if you look up to the Moon one night with a little grain of salt in your hand, or your mind. The grain of salt (our reduced Earth) remains the same. But the blackness, the universe, now extends 225,000 miles to the left of your grain of salt, to the right, the same distance up and... well, my, that's a long way down!

Now this is big, big. And, just about possible to imagine as the Moon gives you a reference point. The trick here is not to look at the Moon as you would normally do, from your 'full size' perspective. You need to feel the distance from the perspective of the grain of salt. Because the salt grain is Earth – all its mountains, seas, deserts, cities – and 'you', at some impossibly small size inside it. Look up from that salt grain to the Moon and then try to imagine the same distance behind, above and below you – a 450,000-mile-diameter black sphere.

Step 5

Now, if you can, imagine this three-dimensional sphere expanding again to embrace Mars. You can see Mars at night with the naked eye. Our little grain of salt, which is still the same size, is now suspended in a sphere of mostly blackness that stretches 34 million miles to the left, right, up, and you really don't want to be looking down.

Now, you need to work at this to try to hold on to the scale. It's not the Earth at normal size in relation to the distance to Mars, it's Earth reduced 42.5 billion times. The grain of salt in your hand suspended in a black sphere that's now 68 million miles in diameter, creating a truly immense, almost overwhelming, volume of space.

Step 6

Now, imagine this sphere of your worst nightmares expanding yet again, further and further out from your little salt grain, past Mars, Jupiter, Saturn, Uranus, Neptune, until it grows so big it encompasses the outer reaches of our solar system, its circumference embracing Pluto. (This is the planet, remember, that NASA's 'New Horizons' probe took nine years to reach, flying at speeds of 52,000 mph.)

Your tiny grain of salt (our planet) is now crushed to oblivion in a monster of a sphere that extends 2.66 billion miles to the right (Pluto at its nearest to Earth), 2.66 billion miles to the left, the same up and, yep, that will be 2.66 billion miles drop. A 5.32-billion-mile-wide monster. And Earth, just a 0.3 mm grain of salt.

If you can follow this mentally, and feel it, it should make you shudder as you get a glimpse of the reality that lurks outside our little patch of blue sky. There are no words in the English dictionary to truly do justice to the size of the thing we are talking about. The thing we exist within. I've worked at it, gone through this little mental process or 'game', and in the briefest of instances I've 'seen it' and it makes me breathless. You almost have to step back from it, it's so unimaginable. But, the problem is, the real unimaginable is yet to come.

Step 7

Get that 5.32-billion-mile-diameter sphere in your head, with all its monstrous volume, because the thing's just going to keep growing.

First, it's going to double. As if 5.32 billion miles across wasn't ridiculous enough, it's going to go to 10 billion miles in diameter. And before you can grasp that, it's going to double again, to 20 billion miles across.

At this stage, its circumference is a colossal 63 billion miles! And while the little grain of salt remains the same in your hand, representing everything that is us, on our Earth, the monster is going to double again to 40 billion miles across, then again to 80 billion, to 160 billion, 320 billion, 640 billion and then a breathless again to 1,280 billion miles! That's beyond a trillion miles in diameter! And all this time, through all these horrific doublings, the grain of salt remains just a grain of salt.

Remember just how a big trillion is. A billion seconds is 30 years; a trillion seconds is 30,000 years! We have a trillion-mile-wide black sphere in relation to us, a speck-like grain of salt. I think that delivers on the idea of unimaginable.

And now for the coup de grâce (not Step 8!).

Step 8

This trillion-mile-diameter sphere is going to double to two trillion miles (stop it, someone!). Then it's going to do it again to four trillion. Then, this monster of all monsters is going to double again, to eight trillion miles. A grain of salt, still in your hand, our Earth suspended in a black sphere with four trillion miles to our left, four trillion to our right, four trillion miles up and (no thank you very much). Is that it? Is this enough? Not quite.

It's going to grow another trillion miles to nine trillion.

Then 10, 11, 12, 13 trillion miles in diameter.

And, finally, it adds one more trillion miles to 'settle' (for the time being) – a 14-trillion-mile-diameter sphere of black, black space.

And we are just a grain of salt.

I think I need a lie down.

Remember, on any given day, when you're shopping, on a train, taking a walk, imagine the grain of salt in your hand. Then look up to the sky and the blue dome it creates over our Earth. Picture the little grain of salt in relation to the volume that dome creates all around you. (I do this a lot, especially in places like a beach or open countryside, where you get a 'big sky'.) Even at this scale the grain of salt is so small it's hard to give it any significance in relation to the volume of blue sky you can see and feel above your head. But you are only looking out to a distance of 62 miles, to the edge of the Earth's atmosphere. To get to the size of the visible universe, at this 'salt size' scale, you would have to multiply that 62-mile volume of sky above your head by an unfathomable 206 BILLION!

Look down at that little salt grain again – at 'us' – now look up, taking in the whole sky. Now times that by 206 billion.

And there you have it.

Fly over the Atlantic Ocean with a hapless little ant dangling from your fingers. Drop the ant down into that vast, vast, seemingly never-ending, deep expanse of blue. It would take a nonillion more Earths to fill the observable universe than it would take ants to fill the entire Atlantic Ocean. That's this many – 1,000,000,000,000,000,000,0 00,000,000,000 – *more* Earths to fill the volume of the visible universe than ants to fill the total volume of the Atlantic Ocean.[9] If you can get any of that in your head. If you can see or feel 'us' in relation to the reality we exist in, this unimaginable behemoth of a thing, it should take your breath away, too. It can also beg questions: who are we, then, and what on Earth are we doing here? It does kind of put into perspective the lives of the 7.6 billion of us who are 'so important', scurrying around on this little salt grain.

Before we finish, we have one last job to do – bring us and the universe back to size.

So, times our little grain of salt by 42.5 billion and here we are, back on planet Earth, full size. Times the 14-trillion-mile sphere by 42.5 billion and you get the full size of the visible universe – 93.4 billion <u>light years</u> across (up-down).

That's 93.4 billion <u>light years</u>! I mean, come on! That's light, which can travel at a ludicrous 671 million miles an hour, and even at that bonkers speed it's going to take 93.4 billion years to go from one side of the visible universe to the other. That's nearly nine times longer than the universe itself has been in existence. That's how big, big is. And that, my infinitesimally small, mind-boggled friends, is just the visible universe. The bit our telescopes can see. There's more beyond that. Possibly a never-ending, infinite more. How big is that?

Well, take the 93.4-billion-light-years diameter of the visible universe and double it. Double it again and again and again and again. And keep on doubling it, for all eternity.

That's roughly how big infinity is.

CHAPTER 4

HOW SMALL IS SMALL?

So, we've done big. At least, to the size of the visible universe, which should sate most people's appetites. And if the reality of the visible universe's enormity hasn't left you a little breathless, then some of the monsters that inhabit this vastness should do the job.

But, before more 'big stuff', let's talk small.

'Small' is more difficult than 'big'. You can see big things, e.g. you're likely to see The Hound coming long before you see Tyrion Lannister. We can see the whole of the Sun. We can see Mars in the night sky. We can even see Andromeda with our naked eyes, all 220,000 light years across and two and a half million light years in distance. Our eyes and our telescopes allow us to do this, to 'see big'.

And even if you can't see something in the visible universe in its totality, it's possible, just, to try to envisage it, imagine it, using an analogy like the grain of salt to conjure it up. That's because we live in the big. We live in a world of tall buildings, mountains, vast oceans and deserts and the Moon and Sun. We see 'big' and we are conscious and aware of it. We don't live and aren't conscious of the world of the 'small' – molecules, bacteria, atoms, and stuff even smaller than atoms. Or, for instance, the *Scydosella musawasensis* beetle, the world's current smallest free-living insect, just 0.325 millimetres 'long' (not sure the Scydosella really qualifies for use of the word 'long').

In our day-to-day experiences these tiny things have no reality. You climb a mountain, you fall off a mountain. You tend not to do this with bacteria. We put men in big rockets and send them to the Moon. No men in rockets will be landing on an atom any day soon. This 'small, small world' (you can hum the tune if you know it) is not intuitive in terms of our everyday experiences, it's not 'real'. (Unless you catch a dose of campylobacter bacteria from an uncooked sausage off a barbie. It feels pretty real then!)

So, small is harder to grasp. But it's important. Because small is what makes up all the big stuff. In fact, searching for the smallest of smalls is actually (drum roll) 'the search for reality itself'. Reality, everything you see, feel, hear, experience, including you, is made up of something. And, as we will come to see in a later chapter, however bizarre it may sound, even space – that seemingly vacuous thing outside of

Earth – and time, which we think of more as a concept than a physical thing, are both made of something. Something terribly, terribly... small.

So, worth a few pages to have a go at getting our heads around it, even if it feels a little less exciting than big. Let's start with the smallest thing we can see comfortably with the human eye: our grain of salt from the previous chapter. (Out of interest, a grain of salt, at around 0.3 millimetres, is roughly the same size as the *Scydosella musawasensis* beetle. Be careful which one you sprinkle on your burger!)

What is it that we can't see, that the grain of salt itself is made up of?

Atoms. The building blocks of us and everything around us.

So how small are atoms?

Well, if you're still holding that tiny grain of salt from Chapter 3 in the palm of your hand, you're holding 1,000,000,000,000,000,000 atoms! (If you read Chapter 1 'Big Numbers', you will know that six groups of three zeros is one quintillion, which was also the number of miles across the diameter of the Milky Way, so you know that is a big number!)

How on Earth can a single grain of salt subdivide a million, million, million times? How can anything that small actually exist? You could stare at that grain of salt forever and never get close to grasping the quintillion individual atoms that have clumped together to make just that single salt grain. (I'd even struggle to imagine it divided into quarters!)

That's the problem with small. Nothing we experience now and nothing in our two million years of human development has prepared us for the 'beyond infinitesimally small' building blocks of existence. But, because we are not just 'laymen', but very much 'boldly laymen', we are going to have a go! We are going to try to unpack that little grain of salt into its individual component atoms, so you get at least a feel for the horrifically small scale that atoms exist at.

Here we go, it may help, it may not. It might make you feel sick.

Picture the little grain of salt in the palm of your hand, all one quintillion atoms' worth. Now imagine breaking it apart into a thousand smaller pieces and laying them all out in front of you. (Stop, shut your eyes, imagine it.) The first shocking reality is these impossible-to-see 'thousandths' of the original salt grain each contain a quadrillion atoms! That's each one. (Picture it, take a breath, move on.)

Now, take just one of these thousand 'quadrillion salty bits' and break just that one into a further thousand pieces. Try to imagine these thousand smaller pieces of salt. Then realise that each one of these thousand, newer, 'incy wincy salty bits' has a trillion atoms inside it. Each one.

You now have in front of you 999 'quadrillion salty bits' from the original subdivision and <u>one</u> of the 'quadrillion salty bits' broken down further into a thousand 'incy wincy salty bits', each with a trillion atoms inside. The second shocking reality here, at least it always is for me, is that you have only subdivided <u>one</u> of the 'quadrillion salty bits' to get to a thousand new 'incy wincy salty bits', each with a trillion atoms in. You have to go on and subdivide the remaining 999 'quadrillion salty bits' still in front of you until all one thousand 'quadrillion salty bits' are subdivided by a thousand, to arrive at one million 'incy wincy salty bits' in front of you. <u>Each</u> with a trillion atoms in it!

Keep going, if you can.

Now take <u>just one</u> of these 'incy wincy salty bits' (just one of the million in front of you) and subdivide it into a further thousand smaller bits. You now have one thousand 'even smaller than incy wincy bits' in front of you, <u>each</u> with a billion atoms inside. And yes, you now have to subdivide the other 999,999 remaining 'incy wincy salty bits' so each of them is broken down into the 'even smaller than incy wincy bits', each with a billion atoms inside.

Where are we? Where are you?!

You are now sitting in front of one billion 'even smaller than incy wincy bits' of salt (which have probably lost their 'saltiness' by now). And each of these billion salty bits still has, quite beyond belief, a billion atoms inside it. Each one!

Tired? Need to sit down?

You can't see any of these billion 'even smaller than incy wincy salty bits' with a billion atoms in each one, of course. But take one, and as you are now used to doing, subdivide just that one into a thousand 'bits of even smaller than incy wincy salty bits'. Each one of these new thousand 'bits of even smaller than incy wincy salty bits' still has a million atoms in it. But, of course, you have to subdivide the other 999,999,999 (nine hundred and ninety-nine million, nine hundred and ninety-nine thousand, nine hundred and ninety-nine) other 'even smaller than incy wincy salty bits', by a thousand.

And when you've done that you have a trillion 'totally invisible salty bits' in front of you. Each one of those trillion 'totally invisible salty bits' contains one million atoms!

We'll go quicker now, in case you've turned to stone.

One trillion 'totally invisible salty bits' in front of you. Subdivide each of the trillion 'totally invisible salty bits' by a thousand and you get one quadrillion bits of 'you can't possibly call that salt!', each still with a thousand atoms inside.

One quadrillion bits of 'you can't possibly call that salt!' in front of you. Subdivide each 'bit' by a thousand into 'where's the salt?' bits, and you arrive at... one quintillion individual atoms! (Hurrah!!)

Just from one 0.3 millimetre grain of salt.

Now you've unpacked your little salt grain (who would believe it would take two pages to unpack one grain of salt?) you might as well count them as you put the grain of salt back together.

Remember in Chapter 1 'Big Numbers', we saw how long it would take to count to a trillion? (Around 542,241 years.) Well, a quintillion is way, way beyond a trillion. To count all the individual quintillion atoms you have just unpacked, from just a single grain of salt, would take in the region of… 542 BILLION years. That, my 'boldly layman friends', is 41 times longer than the universe itself has been in existence. Just to count the atoms in a grain of salt! Now, come on… that alone is worth buying the book for, isn't it? I put the words down on the page and still I need to read it over and over to 'get it'. And each time, I read it as if it were new, such is the wonder of it.

And just think, the *Scydosella musawasensis* beetle, at 0.325 millimetres, must weigh in at around one quintillion atoms. Just imagine you say to a friend, 'See that Scydosella there? I'll give you a million pounds if you can count all the atoms he's made of.'

Your friend goes, 'A million pounds? Just to count the atoms in that tiny thing? You bet!'

542 billion years later… ha, ha, ha!!!

Don't you just love it?

Here are some other ways to grasp or be completely dumbfounded by these impossibly small, but also impossibly important, things called atoms.

Look in front of you. Look at the 'invisible' air. Wave your hand through it. Nothing there, right? Well, with every breath you inhale while reading this little chapter, you breathe in around one litre of air. And with that single breath, that litre of air contains around 1,000,000,000,0 00,000,000,000,000 atoms! That's one septillion atoms! (That's a thousand sextillion, which is, itself, a thousand quintillion!) Take a look back at the air in front of your face. Concentrate, take a deeper look. Where, exactly, are those septillion atoms you're about to take in, in just one breath? You have to hand it to the atom, it really does know how to do 'small'.

Let's take something else, equally close to hand: the paper this book is printed on. Turn the edge of the page towards you. Look at the sliver of thinness that is the paper's edge. Atoms are so small you could line up around five million 'average atoms' and they would just about stretch across the thickness of this paper's edge. (I say 'average' because the hundred-plus types of atom in existence come in slightly different sizes.) Stare at that edge again. Even trying to imagine a hundred atoms across its thickness is impossible. But it's not a hundred, it's not a thousand, or a million – it's five million!

I'm sure you're done with analogies. You get it, they're bloody small. But I particularly like this. It's by a very clever man called John D. Norton and this is his analogy to help us understand just how small atoms are.[10] He asks us to imagine we are offered one atom of gold for every second that has passed since the start of existence. That's every second that has ticked by since the Big Bang, the beginning of time. That's 13.8 billion years! (I, for one, would take the deal; I'd sell the family heirlooms, the wife, the kids... just think about the gold!) The big day arrives, the family's gone, but who cares! You're about to get your gold! (Rubs hands in a Fagin-like manner.) What do you get? A tiny speck of gold weighing all of 0.14 milligrams. (There's a billion times more gold in my wife's wedding ring – where is she?!) The value of this tiny speck – around 0.15 pence! (Or 0.21 cents/0.0017 Euros.)

That brilliant description really grabs me. The number of seconds that have elapsed since the time of creation, the beginning of time, is – give or take a few seconds – 432,000,000,000,000,000. That then equates to 432 quadrillion atoms of gold (and you know how bonkers-big a quadrillion is) and what do you get? Just about enough to have a tooth crowned.

One last atom example. The following are millimetres, |||||| the same millimetres you find on a ruler. Within just one of these millimetre lengths you could line up, side by side, around 10 million atoms.[11] Take a close look at the space between two of these little lines again and then try to imagine anything so small that you could get 10 million of them sitting side by side in that space. (See the full stop you just passed, at the end of the last sentence? It contains around five million atoms.)

Given all of this, you would have thought we were done with 'small'. But, oh no. Just as the grain of salt is built out of atoms, so the atoms themselves, these invisible bits of seeming nothingness, are built out of other things! Other very, very small things. And if we can't truly grasp the smallness of atoms, then we're going to have very little chance of grasping any of the following. But, as this book is for 'triers', let's go.

Inside each atom, at its centre, is a nucleus. So, another 'thing'. And if you think the atom did a good job at being small, the nucleus puts it to shame. Just a phoney masquerading as small. Because the nucleus of an average atom is about ten thousand times smaller than the atom itself... and takes up only a trillionth of an atom's volume! Think about this sheet of paper again, needing five million atoms lined up side by side to cover the thickness of its edge. Now make that the nucleus, inside the atom, all lined up across the paper's edge and you would need an astonishing 50 billion of them!!

If you stare at that sliver of an edge, it's truly mind-shredding to believe anything could be that small and still have any claim to existence; that is, have any physical reality. How did we ever find stuff like this? Even more astonishing is the fact that we have learnt to manipulate things this small, for good and, as we will see, for very bad.

One more thing about the impossibly small nucleus – it's ridiculously dense (like spam is dense and soup is not). This feels weird, given we've just seen that the nucleus of an atom hasn't any real size at all, occupying a volume of space that is close to zero. But dense, dense, deadly dense it is. How to get a handle on this quite monstrous density?

First, conjure up an image of a 1 ft square box. Now imagine we get a super-duper car-crushing machine and crush six billion cars down to a size where all their metal, electronics, wheels, interior, and everything else in all six billion of them fits inside our box. Imagine the density of all those six billion cars forced into such a confined space. Also, think about their weight. With each car averaging two tonnes, our box now weighs a staggering 12 billion tonnes, with the crushed metal of six billion cars packed inside it.

That's how dense the nucleus is inside an atom. As dense as a 1 ft square box weighing 12 billion tonnes, with the crushed metal of six billion cars packed inside it. While the nucleus is infinitesimally small, occupying only a trillionth of an atom's volume, it carries this kind of density. And, as we will see, this gives the nucleus the deadliest potency.

So, we've reached the nucleus of atoms. In terms of our little ruler picture with all its little millimetres, if we needed ten million atoms to cover the length of one millimetre, we need an unbelievable 100 billion nuclei! Look at that little space, please. |||||| How impossibly small, while still being real, can you get? The answer, you know, is yet smaller. Because the nucleus is not it. The nucleus itself is made of two other (how do you say 'really tiny?') really, really tiny things, or particles, as they are more properly referred to. 'Ladies and gentlemen, a big round of applause, please, and welcome to the proton and neutron!' I mean, come on, how much smaller than the laughably small nucleus can these things be?

OK, get your boss-eyed stare back to our little millimetre picture above. You would need one trillion neutrons or protons lined up side by side just to cover that one-millimetre length! That's a thousand billion protons or neutrons (or a combination of both, I don't mind!).

You've got a quintillion atoms in a grain of salt. Each of these individual atoms has a nucleus, which on average is ten thousand times smaller than the atom it lies at the centre of. And then, lo and behold, this nucleus has not one, but two, bits of even smaller stuff inside it! It shouldn't be allowed! How can anything be so small that you would need a thousand billion (one trillion) of them lined up next to each other just to cover the length of one millimetre? And, I repeat, quite astonishingly, 'we' (be sure to take the credit for all the amazing things man and womankind have achieved so far in our two million years), yes 'we' have created technology to actually measure these things and use them to our benefit.

So, are we there?

Not quite.

We need to take one final 'small step for layman-kind'. The proton and neutron don't have it as kings and queens of small. They're not the real deal. They, too, have structure. Quite unbelievably, they are both made of something. And that something is a little fellow called a quark. And if the title of 'smallest' wasn't enough, the quarks have the cheek to be so incy, tincy, bincy, wincy and any other 'incy' you want to add, that three of them can cram inside a neutron particle and three inside a proton particle. Three of the little buggers!

And, would you believe, they also come in 'up' and 'down' varieties! (The proton has two 'up' quarks and one 'down', while the neutron has two 'down' quarks and one 'up'.) So, my weary band of 'small' followers, how small exactly are these funny little head-standing Disneyesque particles?

Well, they are so small that 'we' (maybe we should forget the collective bit here and blame the scientists), that 'they' have yet to find a way to give quarks any size description at all! No actual measurement. Occupying no actual volume!

Now that's what I call 'small'!!

'Hello, meet my husband Donald.'
  'Eh... I'd love to but, excuse me, where is he?'

'Oh, he's here all right, it's just Donald's so small, you know, that you, me, even the kids can't see him.'

'Not even the kids? How small is Donald, to be exact?'

'Well, he hasn't actually got a size, not one that any of the doctors can measure or even describe... you know, he's really, very small.'

'Wow... I mean, that's tough.'

'No, he's fine with it, you know. Of course, he has his up and down days.'

'Yes, I'm sure... look, I'm sorry, bit of an awkward question... but how do you know your husband, Donald, actually exists?'

Computer models, simulations, mathematics, the Large Hadron Collider (LHC) – these things tell us that quarks exist. The same proven process that tells us atoms exist. Are quarks important? You betcha, because as of this day, they are one of the smallest known building blocks of all matter. Think of anything you can, with any kind of physical presence, and right at the root of what it's made of, you'll find the quark. (Just for reference, the quarks actually come in six varieties, called up, down, top, strange, bottom and charm. No size, no volume, but such wonderful names!)

But let's not leave the little quarks with 'we can't say how small they are'. We do know one thing. If quarks were <u>larger</u> than ten thousand times smaller than a proton or neutron, we would be able to measure them. That means they are <u>smaller</u> than ten thousand times smaller than the proton or neutron. (Some estimates put them close to 60,000 times smaller. Can you imagine?) Using only the size of ten thousand times smaller than the proton or neutron, we can complete our 'millimetre challenge'. How many quarks, one of the building blocks of you, me and everything, would it take lined up in a row to stretch across that vast canyon that is the length of a millimetre?

Ten (ridiculous) quadrillion! (That's the minimum; it's likely to be many more, so double ridiculous!)

For all of you who know your atoms, as we were diving down, you'll have spotted that we bypassed the electron.

Electrons spin around the atomic nucleus in up to nine different orbits, which, in combination, make up an atom's outer shell. They are also the reason atoms can bind together to form molecules. (Atoms either exchange electrons or share an electron to create this binding effect.) In terms of size, the electron (which, if physicists were up for it we could rename 'Donald') has no discernible size, just like the quark. And, at present, both the quark and the electron are called 'fundamental particles', meaning we don't think they are actually made of anything else. They, in terms of matter, are it. At your core, they are you. Or, more accurately, you are them.

However, physicists and super-brainboxes around the world still don't think zero size measurement and zero size volume is small enough (don't you just love them!). They are still hunting for a more 'fundamental' answer to the building block of all things, something that even underpins quarks and electrons. There are theories being pursued about an even smaller particle, called a 'preon'. Others are focusing on almost infinitely small 'vibrating strings' (string theory), and so on. Both things are as yet unproven. Imagine life lived at this level, a day job searching for things that are so impossibly small they make our grain of salt look positively enormous. What's truly miraculous about these 'so-tiny-they-are-impossible-to-conceive-of' particles is that, somehow, they have the ability to find each other and clump together to create what we experience and call matter.

The quarks, with a little help from a Trekkie-sounding particle called a gluon, clump together to create the protons and neutrons, which clump together to create the nucleus, which – together with the hard-working electron – makes atoms. Which, in turn, clump together in an amazing array of combinations to make molecules, which clump together to make a grain of salt, peas, toothpaste, water, you, me, propane gas for your barbie, hedgehogs, planet Earth and a googol number of stars. (It's actually more like a septillion stars, but I like the word 'googol'.) All this from things so small we can't even see them. And some, we don't even have the technology to measure. Given we had the technology to put a man on the Moon some fifty years ago, that tells you something about the impossibly small scale at which these particles exist.

Thankfully they do exist, which means you do. We all do.

The next time someone tells you they don't understand you, help them out and say, 'Actually, I'm a googolplex of quarks manifest as a human being.' That should do the trick. Now, before we move on from the building blocks of life and all things, there's a word we need to embrace and try to come to terms with because it will be important later.

That word is 'mass'.

Mass is a strange thing, a potential rabbit hole, but worth a peek inside.

The tiny particles that make up matter, like quarks, have a physical presence. While they may be at a size that we find impossible to comprehend, they do exist, they have 'substance'. If they didn't have substance, a physical reality, they wouldn't be able to come together to create rock, water, iron, pancakes, you or me. This physical presence that particles have, their substance, is referred to as their 'mass'. (You have a physical presence, and so you have a mass – think of your BMI, your body mass index.)

Different particles have different amounts of mass, some more, some less. Your mass is merely a combination of all the tiny little particles that make you up, and these particles' differing levels of mass. So, hopefully, reasonably simple. And the way we tend to measure the mass of an object is to weigh it. But while that is what we do, it's not an accurate measure.

Take my friend, Kev. Weigh him on Earth and you get a mighty 20 stone. Fine, sounds accurate (he is a big boy). But weigh him on the Moon and you would get around 16 stone. Oh… not so fine, not so accurate. Why? Because there is less gravity on the Moon than there is on Earth, so things weigh less (more on gravity in the next chapter). But, even though my friend Kev's weight measures differently on Earth versus the Moon, his mass doesn't change. He is still the same 'Kev', still made up of all the same stuff. His particle count or composition hasn't changed, so his mass doesn't change.

Weighing something to determine the amount of mass present is not an accurate measure of how much mass an object has. It's only because 99.9 per cent (recurring) of everything we do happens on Earth (not the Moon) that we use 'weight' as a shorthand for 'mass'. But while it's a handy shorthand, it's not accurate. Weight and mass are different. So that's one little thing cleared up. And my mate, Kev, knows there's a different life waiting for him somewhere in the future, where, without any dieting, he can be a svelte 16 stone or less.

Now, that may or may not have been interesting. But what *is* really interesting is this thing called mass is interchangeable (given the right conditions) with energy. In fact, mass and energy are one and the same thing, just in different forms. The simplest layman's analogy I know to help you get this is water and ice. (It's not perfect, but it helps me.)

Water, in the right conditions, is fluid. It flows. It can actually travel at amazing speeds (think of a tsunami, which can reach the speed of a jetliner, circa 550 mph/885 kph).

However, change the conditions (temperature below 0 degrees Celsius/32 degrees Fahrenheit) and it changes to solid ice. Not just ice cubes to plop into your G&T, but huge mountains of ice, solid enough to split the hull of an unsinkable ship. The thing we call mass, the substance of tiny particles, kind of does the same thing. Under the right conditions it can transition from the solid form of mass that makes up matter into pure energy (called 'energy mass'). And, again, given the right conditions, that energy can transition or 'condense' back into solid mass. It's an astonishing thing. The fundamental substance of all things is like a chameleon.

Why is this so important?

Because the atom bomb was built based on the understanding that you can transform some of the insanely dense solid mass that exists in the nucleus of an atom into energy mass if you split the nucleus of that atom. Remember the nucleus? Ridiculously small. Equally ridiculously dense (the same density as a 1ft square box with 6 billion cars crushed inside it). Stimulate this dense solid mass to transition into energy mass, and the energy release is, regrettably, imaginable – the Hiroshima and Nagasaki atomic bombs.

The Hiroshima atomic bomb was dropped on 6 August 1945 and used a chemical element called uranium. The

Nagasaki bomb was dropped on 9 August 1945 and used plutonium. Both these atomic bombs only converted about three per cent of the potential energy of the atomic nucleus that was split. But the splitting of each initial nucleus caused other nuclei within the uranium and plutonium to split as well, creating what is known as a chain reaction. The result was the energy equivalent of 15,000 tons of TNT exploding over Hiroshima and 21,000 tons of TNT exploding over Nagasaki, with temperatures at the core of both explosions reaching in excess of ten times that of the Sun, some 200 million degrees centigrade. These two events gave physical reality to the staggering density of the atomic nucleus and its latent energy in the most horrific of ways.

That's why this curious word 'mass' is so important. Important enough to lie at the heart of the world's most famous equation: $E = mc^2$. It was Albert Einstein who first realised that mass and energy are one and the same thing. A 'thing' that can manifest in different guises.

And it was an equally gifted scientist, Ernest Rutherford, who discovered not only the nucleus that lies at the centre of atoms, but also conducted the first experiment that split the nucleus of an atom (called nuclear fission). The work of both Einstein and Rutherford came together in the Manhattan Project in 1945. And our world changed forever.

Anything made of matter has the energy potential of the Hiroshima and Nagasaki atomic bombs, because all matter is made of atoms. Were you able to convert all the solid mass in a paper clip into its energy equivalent, you would, quite unbelievably, release the same power as the atomic bomb dropped on Hiroshima. Just from a single paper clip. You and I have this energy potential. If you could convert the solid mass of an average-sized adult into energy mass, it would be the equivalent of around thirty Hiroshima bombs. Thankfully, we don't possess the capability to release the energy locked up in you, me, paper clips or other forms of familiar matter, and are unlikely to ever do so. However, we have continued to find ways to unlock more energy from the nucleus of certain atoms.

In 1961, on the 30 October, Russia detonated their RDS-220 hydrogen bomb, nicknamed the 'Tsar Bomba'. This is still the largest ever recorded nuclear explosion and was calculated to have created an energy release 3,800 times greater than the bomb that exploded over Hiroshima. So, when we think about the tiniest, tiniest particles that make up what we see and understand as our physical existence, don't be fooled into thinking 'tiny' = 'weeny'. These particles have mass. These particles are potential 'Goliaths', deadly dense and full of locked-in energy. We release it at our peril.

So there you have it – 'boring old small'. Everything is made of atoms. Atoms are made up of electrons, spinning in orbits around the tiny, dense nucleus. The nucleus is made of protons and neutrons, and these are made of the little upside-down 'jokers in the pack', the quarks, held together by 'a squeeze of gluon'. And the quarks, along with the electrons, are it. They don't subdivide, as far as we can tell. Everything we know, including us, will ultimately decompose down to these fundamental little particles.

They are, quite prophetically, the source of life and at the same time the potential source of Armageddon.

CHAPTER 5

SPACE AND TIME

So we've done big numbers and big distances. We've gone to the edge of the visible universe and done an *Alice in Wonderland* dive into the smallest things that make up most of what we see in the visible universe. But what about space itself? This thing that fills the visible universe and which everything else, including us, 'sits in'. Well, space is something the majority of us take completely for granted; after all, it's just… space. How wrong could we be?

First things first. Despite how we might think about space (space normally means there's nothing there) and the language we tend to use to describe space (empty), space is not empty and it's definitely not 'nothing'. Space, that dark black stuff that Neil Armstrong, Buzz Aldrin and Michael Collins flew 225,622 miles through to get to the Moon, is an entity, a definite 'something'. (How can you fly in a rocket through nothing? There wouldn't be anything to fly through!)

Our intuitive reaction to space belies the true nature of space. So, where did this something called space come from? And what is it made of? Well, to all intents and purposes, space was born around 13.8 billion years ago in an event that is referred to as the Big Bang. The thing to recognise here is that before the Big Bang, the three-dimensional space we are now living in didn't exist.

'Our space' came into existence 13.8 billion years ago, born from a microscopic dot, millions of times smaller than a quark, which then proceeded to mushroom outward in all directions at a truly frightening rate. A dot of possibly infinite density, in the region of four to five trillion degrees Celsius (that's around 300,000 times hotter than the centre of the Sun!), which grew and grew and grew, to the most horrifically, humongous size it's possible to imagine, 93.4 billion light years across, and then some.

Now that's hardly nothing.

So, where did the microscopic dot that ultimately gave birth to us come from? We don't know. What did the microscopic dot exist in? (Surely, even a dot smaller than a quark has to exist 'in' something?) And this perhaps is the weirdest of thoughts. We don't know what the dot existed in. It wasn't in what we experience and understand as space. The space you exist in while reading this book and the vast interstellar space that makes up the visible universe, was in the microscopic dot. It was the dot (your ultimate, ultimate birth mother... the 'dot'). Without the dot there would have been no 'you'.

This is one of the hardest things I find to get my head around. To imagine that 'just before' 13.8 billion years ago, the three-dimensional space we exist in, see all around us and truly take for granted, did not exist. It hadn't happened.

There was no space.

The super-dense, overheated, trillions-of-times-smaller-than-a-pinprick dot was the beginning, start, birth, of the seemingly never-ending space we see all around us. And, equally astonishingly, after 13.8 billion years of growth (which in itself is pretty incomprehensible), the damn stuff is not only continuing to expand, it's also speeding up! We're not just talking Usain Bolt fast here, we're talking speeds up to and possibly in excess of 671 million miles an hour, which means it's accelerating faster than the speed of light! That's another 671 million more miles of space in all directions with every hour that passes!

Now, forgive me for appearing a little assertive here, with all my newly found information, but 'nothing' doesn't come into existence. 'Nothing' tends not to grow. And, as far as I can recall, I've never witnessed 'nothing' accelerate.

And there's more.

When space gets close to matter, like our planet Earth or a star like our Sun, it reacts. It bends, twists and warps. The space around Earth is slightly curved, inwards and downwards, because of the presence of Earth. Space feels and reacts to Earth's presence. The closer down to Earth (i.e. closer to all the particles that make up planet Earth) the more space bends and curves. Similarly, space will straighten or 'flatten out' the further it is away from Earth, until it's at a distance that the effect is so negligible as to have no effect.

The more matter there is, say in the case of a big planet like Jupiter, the more space twists, bends and curves downwards. Look at something even bigger, like our Sun, and you will see the space around it curve and bend even more. The reason Earth circles around the Sun, some 93 million miles out, is because Earth is following the curved and downward shape of the space around the Sun. Pluto does the same thing, some four billion miles away. That's how far out space curves as a reaction to the Sun's massive size.

And it's not just about the size of the object. Some stars can be relatively small, but incredibly dense – lots of matter compressed down and packed in tightly, giving them extreme levels of mass (think my mate, Kev). Space reacts not just to the size of an object, but to how much matter is compressed inside that object. This warping and bending of space in and down towards an object it 'feels' is what we call, and experience as, gravity. Gravity is just a name we have given to describe the way space curves and warps in the presence of matter. It's why things fall to Earth. If the space around Earth wasn't curved downwards, objects wouldn't fall, they would float. They fall because they have no choice but to follow the inward and downward curving of space towards

Earth, just as Earth has no choice but to follow the inward and downward curving of the space around the Sun.

The only reason we don't fall straight down into the Sun is that Earth has just enough angular momentum (think speed) to resist the downward curving of space, keeping us at a nearly constant distance out from the Sun. But it takes some speed to do that, a cool 67,000 mph (107,826 kph). It's amazing we have no sensation at all of Earth travelling through space at this speed, close on 19 miles a second. (Keep your fingers crossed nothing comes the other way!) The International Space Station (ISS) also needs a constant speed of just over 17,000 mph (27,358 kph) to keep it circling around Earth. Without this speed/angular momentum both us, Earth, and the ISS would most definitely drop.

When a rocket sets out to leave Earth, it's fighting against this downward curvature of space. The rocket has to create enough thrust and get up enough speed to force itself upwards. This speed is called 'Earth's escape velocity' and it's 25,000 mph (40,000 kph). If that speed isn't reached, the rocket will succumb to the downward pressure of the curved space and fall back down to Earth. The further a rocket gets away from Earth, the less acceleration it will need, to the point where the dragging down effect of curved space is so small that the rocket needs no acceleration at all to prevent it falling back to Earth. (It's worth noting that while the effect of curved space becomes increasingly weaker the further you are, say, from Earth, the effect is never zero. Somehow, curved space around any object like Earth has an impact throughout the visible universe. There is no point or distance in space, however far away, in which the curvature is not present, however minuscule it may be.)

Does any of this affect us? The fact space twists and turns in the presence of matter? It's unlikely any of us is going up in a rocket anytime soon. Well, the answer is yes, most definitely. The only reason your feet are planted just perfectly on the surface of the Earth – not so light that you float occasionally, not so heavy you can't lift one foot in front of the other – is because of the nature of the curved space you – 'we' – exist in. If it was any less curved, we literally wouldn't be able to keep our feet on the ground. So we would likely have evolved as a completely different species, or maybe not evolved at all. If it were more curved, we would be flattened to the face of the Earth like a cowpat. We just happen to exist on a little round ball of matter that causes space to curve at just the right amount to allow us to exist. The next time you drop your mobile, fall over, or a bird empties the contents of its bowels on your head, go, 'Yeah! Curved space!'

So, Earth is an example of how matter causes space (this 'nothingness') to curve, exampled by how gently our feet connect to the planet and apples fall rather than float. But there's matter (mass) out there that gets served up in volumes millions, billions and trillions of times more than that of Earth. And when space feels these monster concentrations of matter it crumples, buckles and folds in on itself to such an extent that in some regions of contracted, twisted space nothing can escape. Not even light. Light travels through space. So if space is all mangled up and turned in on itself, light can't help but follow this twisted, downward path. And it does. Until it gets trapped, like a contortionist in a box. And when light can no longer travel we, of course, can't see it, making that intense, twisted region of space totally black. It's regions of space like this that we refer to as black holes (more on these gruesome apparitions in Chapter 7).

To try to get a feel for this more extreme twisted space effect, open your hand so your fingers are stretched out and your palm is flat. Now imagine I place a tiny ball of light in the centre of your palm, just pea-size. The light beams from this pea-size ball can travel across your palm and beyond, lighting up all the stuff they travel to and touch. Now, close your fist as tight as you possibly can, every finger squeezing inwards until they go red. Not only can you not see the little ball of light, the light beams from the little ball can't travel. They are locked in, trapped by the crushing power of your fist. Whilst not perfect, this at least gives you a sense of how space reacts to very high concentrations of matter. It acts like a cosmic fist, tightening and tightening and twisting and turning until it's so contorted nothing within it can escape its crushing effect. Not even light.

Now, I ask you, have you ever seen nothing do that? Methinks not. So, 'Case made, m'lord,' – space is definitely a something, not a nothing. Well, nearly. What about this 'empty' business?

Quantum foam. These are two words quantum physicists use to describe what space itself is made up of. What we see as a black flatness of 'nothing' is predicted to be made up of a teeming, seething sea of impossible to see, feel or even imagine (and I quote), 'Dimensions that unfurl, then furl back in on themselves, spontaneously appearing and disappearing with inconceivable quickness. They blink in and out of existence like the bubbles in the foam of a freshly poured beer. There's no such thing as empty space, there is only "quantum foam", everywhere.'[12] I couldn't have put it better myself. So I didn't try. (I am only a 'boldly layman', after all.)

What this quote suggests, and what particle physicists are beavering away at, is that space itself is, at its deepest roots, granular. That if you were able to break space down to its fundamental constituent parts, you would find the thing it's made of. Just like physicists have broken matter down to find its fundamental constituent parts: quarks and electrons.

And what do the theories suggest the building blocks of space are?

Tiny 'loops'. And these little fellas take 'small' to the ultimate level. One millionth, of a billionth, of a billionth, of a billionth of a centimetre! Small enough? Well, it has to be, because at this size ('size' really isn't the right word), they are thought to represent the smallest bit of reality that can actually exist. Beyond this you truly do get nothing. (Whatever that looks like.) This 'size' is called 'the Planck length' after a very brilliant and famous German physicist called Max Planck.

And it is generally agreed and accepted that nothing can exist beyond this size. At least, not in our universe.

So, what do these little loops do? The theory suggests that these loops link together, or overlap. And where they connect or touch (called 'nodes') they create a tiny piece of three-dimensional volume. Think about a sexagintillion of these little loops (that's '1' followed by 183 noughts) interconnecting a sexagintillion number of times and you create a volume of three-dimensional space, 93.4 billion light years across – the visible universe. And remember, these little loops don't exist 'in space' – like you, me, the impossible-to-measure quark – they are the very fabric of space itself. As you can see, space is not empty. It's full up with quadzillions upon quadzillions of little things doing the 'loop de loop', creating the three-dimensional space that we exist within.

So, this black stuff, space.

It was born. But we have no idea within what. It's grown to an almost unimaginable size. And, after 13.8 billion years, it still hasn't stopped. It's actually continuing to grow and, as bonkers as it may seem, it's accelerating (over a million more miles in all directions in the six seconds it took you to read this sentence). This space stuff can also shape itself – warp, bend, twist and tie itself up in impossible knots. And, though our natural tendency is to think of space as empty (hence our choice of the word 'space'), it's not. Space is a 'thing', it's made of a 'thing', albeit quantum foamy dimensionally loopy things that exist at a size we are never likely to be able to see or even comprehend. And, just when you think the case for space not being nothing and not being empty is as good as won, there's the final twist.

'Time, gentlemen, please.'

Of all space's peculiar characteristics, this is the *most* peculiar. When space curves and warps because it feels the presence of matter, it causes time to change with it. Because space and time are inextricably linked; they are bound together, peas in a pod, part and parcel, as one. It's why they are referred to as a collective, 'space-time'. And when time changes, as space changes, it does the weirdest thing. It slows down.

However incomprehensible this may sound, it's true. It happens. When space curves and warps, it changes time. Time 'ticks' slower, and so passes slower. The more curved, contorted and twisted the space, the slower time passes, to such an extent that when space ties itself up into an impossibly tight knot due to the presence of matter that is incredibly dense and concentrated (like the matter in a black hole) time, as we know and understand it, comes to a virtual stop.

For the majority of us, and I include myself here, it's an impossible thought to truly grasp. What on Earth does it mean to think of time stopping? This goes against all our intuitions and most definitely our everyday experiences. We think of time as consistent, a kind of universal clock that ticks along without variation. Well, time isn't consistent, and it doesn't 'tick along' at one rate. Time is affected and is changed by the curvature of space. If space is very gently curved, like the space at the surface of the Earth, time will tick along at the rate we have become used to. But if it's in a region of space that's more curved, it will tick slower. More curved again, slower still. Curved to the point where space is folded in on itself, screwed up into an irreversible knot... time stops. There's no time as we know it. There is no 'ticking'.

Mind-boggling? Yes. But here's something even more bonkers:

If you were living in that extreme curved space, where time is ticking really slowly, you would age slower in relation to a friend of the same age who is living in space that isn't so curved. And if that wasn't hard enough to come to terms with, both you and your friend would think time was passing quite normally. Everything about you, the way your biology functions, even the way your mind functions, would 'tick' in harmony with the space you were in. It's only if you and your friend were to come together after 'some period of time' living in your respective curved and less-curved regions of space, that you would find you had, somehow, remained miraculously young, while your friend in the less-curved space had aged a year, or ten, or maybe even fifty, more than you. Or, your friend might even have died thousands of years ago, depending on just how curved the space you lived in was. (It's worth mentioning at this point, we have no way of getting to regions of space that are so curved as to significantly make a difference to the way we age. It's also unlikely we would survive. Because space that is that contorted would bend and twist and crush our fragile little bones to smithereens.)

This effect of time slowing as the curvature of space increases (called gravitational time dilation) is not just something that happens in the far reaches of outer space. It happens here, on Earth. We don't need to travel to age a little slower.

As space feels the presence of Earth (all the particles in the matter that make up Earth) it begins to curve downwards. The closer it gets to the surface of the Earth, the stronger the downward curvature becomes. So, just outside the Earth's atmosphere, space only gently curves downwards, while at the top of Mount Everest it's curved a little more. At the surface of Earth it's more curved again, with the greatest degree of curvature happening right at the Earth's core, or centre. And all the time space is curving downwards towards Earth, time is changing with it, and so 'ticking slower'.

Take two atomic clocks (these things are insanely accurate, they don't 'tick wrong'). Place one atop Mount Everest and the other one on the surface of Earth. What do you see measured from Earth? The clocks measure different elapsed times. The clock on Earth's surface (where space is more curved) ticks slightly slower than the one at the top of Mount Everest (where space is slightly less curved). Why? Because all the individual particles that make up each clock react to the nature of the curved space they are in. The more curved the space, the slower all the particles 'tick'. The difference between the clocks is only nanoseconds. But it's real.

If you spent your life living in a high-rise flat or apartment, on the 20th floor, say, would you age fractionally faster than someone living on the ground floor? Unbelievably, yes. Even your head is ageing slightly faster than your feet, because each part of you occupies a slightly more or less increased level of curvature of space and the particles that make up 'you' (your body) tick in accordance with the bit of space they are in.

For us, the effect is so insignificant as to have no real meaning (a nanosecond is only one thousand-millionth of a second). But when we think about what's out there, be sure, there are millions and billions of places where the effect of curved space creates a situation where time as we know and experience it doesn't exist. In some places, there is no ticking. Imagine other 'beings' that might emerge in regions of space where time ticks completely differently to ours? To us they might appear to have a lifespan of thousands of years.

If that isn't mind-boggling enough, space, and as a consequence time, are also affected by objects moving at speed.

The energies created by a fast-moving object, say a rocket, cause space to 'shrink' in the rocket's direction of travel. That is, the space in front of the rocket bunches up or 'gets shorter'.

Think of a blind on a window when it's down at its full length. Pull the strings so the blind goes up and it concertinas into a shorter length (all scrunched up at the top of the window frame). Bizarrely, this is what space does in front of an object that's moving really fast. It concertinas. Which means (if you can possibly get your head around this) there is 'less space'.

Space actually contracts, meaning a fast-moving object has less space to fly through! Come on, how unbelievable is that? The faster and faster you go, the less actual space there is to travel through. So the total length of the journey between your starting point and a very far away destination in space gets shorter, depending on how fast you accelerate towards it. It's true when you're on a plane, a train, even when you're walking versus sitting. It's just that these kinds of speeds, the ones we experience, are so 'slower-than-the-slowest-snail-you've-ever-seen' that the effect on the space in front of us is infinitesimal, so we are never going to be aware of it.

But if we could ever reach speeds representing a reasonable percentage of the speed of light, space would contract in front of us, creating less actual space (and so distance) to fly through – ponder that, my mind-boggled laypeople! Again, the astronauts would have no sensation or awareness of this. It's us on Earth, not accelerating but just tracking the rocket, who would measure space shrinking. And if that wasn't mad enough, the actual object moving at speed, in this case the rocket, contracts in length as well! However ridiculous it sounds, the faster the rocket goes, the shorter its length becomes (its width and height remain the same).

If the rocket were ever to reach the speed of light, it would contract to a point where it had no discernible length at all! Including the astronauts inside it! Bizarrely, the astronauts in the rocket would still have no sensation or awareness of this 'contracting' going on. They would measure everything on the rocket, including themselves, as perfectly normal. Because all the particles that make up our astronauts, and all the particles that make up their rocket, are moving in perfect unison, 'as one'. It's only us on Earth that would see and measure these length changes happening as we are not, by comparison, experiencing accelerated motion.

And, while all this length and space contraction is going on, you would expect, of course, time to be affected. And it is. A bit more clock proof:

Take two more atomic clocks. Put one on the floor by your feet. Put the other on a 747 and fly it across the Atlantic. When the 747 lands, compare the two 'impossible-to-tick-inaccurately' clocks. The one on the 747 will show a minuscule fraction of a second time difference. It would have ticked slower while travelling at speed (even allowing for the fact its height put it in slightly less curved space). Even worse, the people on the plane would have aged a fraction less than you! However, before you dive for a mirror and start checking for wrinkles, just as the clocks are different by only a tiny, tiny amount (and only atomic clocks can measure such a small difference; it's beyond your common clock or watch) so is the time saved by passengers on the plane. They have only aged less than you by 0.0000001 seconds. (Imagine if they had paid Club Class just for that!) This is because the circa 550 mph speed of a 747 is not significant in terms of shrinking objects or space, and so impacting time. Or creating 'time dilation', as it's more properly referred to.

But what about the pilot of our 747, as he or she is flying all the time? Well, even a lifetime's career of flying long-haul on a 747 isn't going to gain our pilot more than a fraction of a second. In terms of ageing, 0.0000000156 of a second, to be exact, over his or her flying lifetime. Don't think they're going to notice and definitely not worth taking up a career in aviation just to knock 0.0000000156 seconds off your age.[13]

An astronaut orbiting Earth in the International Space Station?

Well, let's say they are up there for a year, travelling at speeds of around 17,000 mph (27,000 kph). When they return to Earth, their clocks will show a 3.8-second difference, compared to clocks on Earth. And, not that you would likely notice it, but they would also have aged 3.8 seconds less than you. Worth it for flying around in circles in the space station for a year? Probably not.[14] So, travelling through space, at speed, causes objects and space to shrink in length, just as matter causes space to stretch and curve. And in both instances the rate at which time passes changes as well. We need to get our heads around this truth, because it's been proved, without a doubt.

However, just as we can't get to, nor would we survive, the extremely curved space where time dramatically slows down, e.g. near a black hole, nor can we get anywhere near the speeds we would need to get to, to dramatically slow time down (yet). For that, we would need to get to speeds that are a significant percentage of the speed of light. If we could one day reach such speeds, what would this super fast travel do to time? And, what we all most want to know, what would it do to ageing? Well, if you could get to 10 per cent of the speed of light, in a super fast future rocket (an unlikely 67 million mph/107 million kph) time would slow for you by around one per cent compared to the time your friends experience on Earth. Your 24 hours of 'fast rocket time' would equate to 24 hours plus 14 minutes of 'Earth time'. While your friends would be absolutely certain that only 24 hours had passed, according to their clocks, they would have actually aged 14 minutes more than you in that 24 hours.

Now, clearly, 14 minutes doesn't sound that significant, but that amounts to 3.5 days each year. And even over an average lifetime of 80 years, your Earth friends would only have aged about a year more than you. No one is going to notice.

So, let's jump up in speed and put you in a rocket for five years, travelling at 90 per cent of the speed of light. That's 603 million mph (970 million kph). Now, for every one day you experience on your rocket (which you think still feels like 24 hours), two and a quarter days pass for your friends on Earth. This means your friends ageing more than twice as fast as you. Neither you nor your friends would experience any difference in the way time were passing. It's only when you meet up, after what you have experienced as five years of travel, that you will see your friends have aged not by five years but by over 11 years (and you have gone back to the future).

Take your rocket to 0.999999 per cent of the speed of light (around 670 million mph – over 1 billion kph) and for every day you experience (perfectly normal days, as far as you are concerned) your friends on Earth would experience two years. Think about this. If you were to be in that rocket travelling at that speed for just one month (30 days), when you landed, your friends would have aged 60 years!

Finally, take your rocket to what science tells us is the physical speed limit, 0.99999999999999 per cent of the speed of light, and every single day that you are on your rocket is equivalent to... wait for it... twenty thousand years on Earth![15]

Your friends aren't going to be there. No one's going to be there. Earth might not be there!!

And you would have just experienced one day.

By the way, in case you're wondering, if you take the train to work every day, does the forward momentum that causes your train and space to shrink in length, also slow time for you? The answer is, amazingly, yes. But for train speeds the time dilation, or slowing of time, is such an impossibly small fraction of a second that you would probably have to travel for a billion years just to knock a minute off your age (and think of the cost of your season ticket!).

What this all amounts to is that there is no such thing as absolute time.

We just happen to experience a form of time we have become familiar with because we inhabit, and are more or less stuck on, a minuscule, insignificant speck of rock in a gargantuan monster of a universe. We've been conditioned to think of time as consistent and independent of everything else. That's just because our experience of the reality of the universe is so stiflingly limited. We are like a goldfish that's born into and lives its life out in a tiny goldfish bowl. It only knows what it experiences and it is conditioned by those experiences. Imagine trying to describe a tsunami to a goldfish!

We're not that much different. The time we experience on Earth is just our version of time. It's all we have ever known. But it's not actually a true reflection of time. And even within our limited experience, here on speck-like Earth, we have measured and so proved the reality of time dilation. Outside of our tiny blue bubble world, time manifests in billions and trillions of different guises. Where space is at its most twisted extreme, at the edge of a black hole, time slows to such an extent that the concept of time as we know it doesn't exist. If we could ever accelerate astronauts to speeds of hundreds of millions of miles an hour, what they would experience as one day might be two days for us on Earth, possibly a month, a year or even thousands of years. But, and it's important to note, the astronauts would not feel they had lived any longer, as time for them would have ticked along as 'just normal'. Time dilation does not give you 'more life' – it just slows your life clock down relative to other people not in such curved space or not experiencing long periods of accelerated motion. It's only when our astronauts return to Earth from their adventures that they would find they had been catapulted hundreds or possibly thousands of years into the future.

A couple of points before we move on. The effect of time dilation has been tested and proven. Has the effect of length contraction for objects moving at speed been tested and proven? There is a difference of opinion in scientific circles. While all the maths 'proves it', some believe it's still not a physical reality. A rocket doesn't actually shrink in length, it's just an effect we observe or perceive. Others argue it's a physical thing. Objects, like rockets or humans, will contract in length under acceleration, like the theories predict. Having read quite a bit about this, my layman's view, for what it's worth, is length contraction is real, a physical reality.

Should we then be fearful for any future astronauts, as our rockets get faster and faster, approaching speeds close to the speed of light? Won't their bodies be contracted to an ultimate 'thinness'? Well, help is at hand, because to keep going faster and faster, we need to generate more energy to accelerate our rocket. But as we pile in more energy, some of it transfers to the mass of the rocket itself (remembering mass and energy are interchangeable) so increasing the mass of the rocket and making it more resistant to acceleration. Like a vicious circle, the more energy we put in, the more the mass of the rocket increases, making it ever more resistant to acceleration. As a consequence, we can never reach or exceed the speed of light, because as our rocket gets closer and closer to light speed, it reaches a point where it acquires an infinite amount of mass. To accelerate it any faster would require an infinite amount of energy. And an infinite amount of energy doesn't exist. So astronauts of the future may be thinned a little, but not thinned to oblivion.

There is a last 'time mind-boggler' to cover. And it involves the speed of light.

The speed of light, you now know, is 186,000 miles per second, and this is the cosmic speed limit. Nothing, maybe other than the expansion of space itself, can go faster. This means any event happening on Earth, on the Moon, on Mars, or even further out in space, will always take the speed of light to reach us. Say you're watching the Olympics live on TV. You're in London and the Olympics is in Tokyo. You think you're watching it live, but the TV signal can only reach you at 186,000 miles a second. It travels from the stadium in Tokyo, up to a satellite orbiting the Earth, down to your TV set and across your living room to your eyes.

Now, at 186,000 miles a second, covering a distance of only around 6,000 miles, you are not going to notice the delay (fractions of a second), but it is there.

Now, imagine we pop another man or woman on the Moon and they broadcast live back to Earth. As you might remember, the Moon is some 225,000 miles away. What you think you're seeing broadcast live actually happened three seconds ago. That's how long it takes the broadcast signal to reach you, travelling at the speed of light.

Move out to the Sun. Say the Sun suddenly explodes. You won't know about it for eight minutes. The Sun is 93 million miles away. It takes light travelling at 186,000 miles per second, eight minutes to cover that 93-million-mile distance. Because nothing can go faster than light, not debris, not sound, not anything, you will have to wait a full eight minutes before you have any idea that the Sun has erupted (and it's not just your holiday to the Maldives that's buggered, it's the Earth!). If you're on the side of the Earth facing the Sun, an electromagnetic tsunami will hit after eight minutes, frying everything you and everyone else in its path holds dear. The physical debris, travelling at a speed considerably less than the speed of light, will hit about 22 hours later, shredding through planet Earth like a cosmic machine gun. The point is, for all you know, the Sun could have exploded four minutes ago, and you have just four minutes from now until oblivion.

This delay in being able to receive information (which is a reality of the physical world) we will simply refer to as 'natural time delay'.

And the weird thing about this 'natural time delay' is you can't tell if an event (like the Sun exploding) is in your past, your present, or still to happen in your future. Think about the Sun. For every eight minutes of your life, you can't say whether the Sun has exploded. For example, if it happened two minutes ago, you will only know about it in another six minutes time, so it's in your past. Or it just happened, as you're reading this line, and you will know about it in eight minutes' time, so it's in your present. Or it is going to happen in your future, in one minute's time, to be exact, and you will know about it in exactly nine minutes. You have no way of knowing until any particular eight minutes has passed. And either the Sun is still in the sky or... goodnight.

That's the Sun. Now go to Mars, where we're planning to visit in the next 15–20 years. Mars can be around 34 million miles from Earth (55 million km) or 249 million miles from Earth (400 million km) depending on each planet's orbit around the Sun. At these two distances, signals from Mars will take either a minimum of four minutes or a maximum of 24 minutes to travel back to Earth. Let's take a 'natural time delay' somewhere in the middle, of 15 minutes.

Let's say on 1 April 2035 a manned rocket is due to touch down on Mars at 9am, Earth time. At the planned time of landing we can't know by any means whether the rocket has landed or not. We will have to wait a full 15 minutes to either receive a successful landing signal or receive no signal. That's a 15-minute window in which it's impossible to receive any information and so impossible for us to know whether the landing event has happened. Again, it could be in our past, present or still in our future. The first indication we will get will be 9:15am, Earth time.

Now, 15 minutes doesn't sound that long.

So let's imagine we are able to send an unmanned spacecraft, a probe, to the nearest known planet outside of our solar system. As we know, it's called Proxima b, and is referred to as 'second Earth'; so an exciting trip. It's 25 trillion miles (40 trillion km) away, or 4.3 light years. Let's say the probe is due to touch down on 25 December 2037. (That's about how long the journey would take if a probe set off now.) The 'natural time delay' is now close to four and a half years. That's how long it will take a signal from the probe to reach us on Earth, travelling at the speed of light. Whilst the probe is due to touch down at Christmas, December 2037, we won't know until around April of 2042 whether it has. Close to four and a half years in which we won't be able to say whether the event of the probe touching down on Proxima b is in our past, present or future. (Out of interest, in 2016 a Russian billionaire by the name of Yuri Milner launched a US$100 million project called 'Breakthrough Starshot', which aims to send a nanocraft – that's small – at some point in the next 20 years or so, to do a fly-past of Proxima b.)

Finally, jump to the next nearest galaxy outside the Milky Way: Andromeda.

Andromeda sits 2.5 million light years from Earth. If there are aliens on Andromeda staring at us now (this minute) through the biggest telescope known to man or alienkind, they would be seeing Earth as it was 2.5 million years ago. The picture they will be viewing at this precise moment will be of our human ancestors, *Homo habilis*, standing around four feet tall, with brains less than half the size of ours, inventing tools out of sharp flakes of stone. (Hence the name *Homo habilis*, which means 'handyman'.) That's because the light images that they are receiving, that travel from Earth to the lens of their telescope, have taken a full 2.5 million years to cross the chasm that separates us and our nearest galaxy. Let's say these aliens are very advanced and rather than just watching Earth, they set off 2.5 million years ago to conquer Earth in spaceships that can travel close to the speed of light.

They could arrive tomorrow.

Because of the cosmic speed limit of light, we now have a natural time delay of 2.5 million years. We ('you') have no way of knowing if that alien invasion fleet from Andromeda set sail some 2.5 million years ago. Their departure from Andromeda could be in your past and they will arrive tomorrow... ALIENS!!! Or they could still be in your future, they haven't left yet. Or they left one million years ago, so have another million and a half years to travel.

In none of these examples can you ever say an event is happening 'now'.

You can't say the Sun is exploding now. You will only know in eight minutes' time, and then it's not now, it actually happened eight minutes ago. You can't say if an astronaut has landed on Mars or Proxima b now. You can only know 15 minutes later, in the case of Mars, or 4.3 years later, in the case of Proxima b. And you can't know what's on its way from Andromeda because of the limit of the speed of light, which determines how fast information can travel, and also how fast objects like spacecraft can travel.

The next time you're in a room or an office and someone calls out to you, remember 'natural time delay'. You think you heard what the other person said instantly, but you didn't. It happened in your past. Only nanoseconds in your past, but definitely in your past. The other person's words still had to cross the space between them and you and can only cover that space at the speed of sound (which is a weedy 767 mph/1,235 kph). You heard the words nanoseconds into the future from when the words were spoken, which was actually in your past.

So, we have two types of time to consider. 'Time dilation' as a result of matter and accelerated motion causing space to curve, and 'natural time delay' due to the finite speed of light, which limits our ability to receive information.

What if we put these together?

Say we were able to send an astronaut to Proxima b, 4.3 light years away, in January 2020. We manage to fly our astronaut there at 40 per cent of the speed of light (268 million mph). At that speed we calculate, based on 'Earth-time', that the journey of 25 trillion miles or 4.3 light years will take our astronaut 10.75 years, taking us from launch, in January 2020, to landing, October 2030. Now, as we know, this kind of speed – 40 per cent of the speed of light – causes space to shrink and time to slow. As far as our astronaut is concerned, the journey to Proxima b only took 9.85 years and he or she landed in November 2029 – an eleven-month difference to the time we have experienced on Earth.

A couple of points here. Both times are real. For us, on Earth, the month our astronaut landed is October 2030. For our astronaut the landing date is November 2029. It's not a miscalculation or a trick, it's the genuine physical impact travelling at these kind of speeds (40 per cent of the speed of light) has on the passing of time. And this time slowing is not just measured by the clocks on board our astronaut's rocket, it's also measured by his or her 'biological clock'. All the atoms which make up our astronaut 'tick' slower, so our astronaut genuinely ages slower, even though he or she experienced time passing just as normal.

So, we already have an eleven-month genuine difference in time between us on Earth, and our astronaut on Proxima b, due to time dilation (And, in comparison to our astronaut, all of us on Earth are eleven months older). We now have to deal with the distance between us and our astronaut, which is a huge 4.3 light years. Look at it first from our astronaut's point of view.

As our astronaut lands in November 2029, remembering that's 'his or her time', the signal automatically fires off from their rocket to inform us on Earth that the landing was successful. This signal can't travel faster than the speed of light, so it is going to take 4.3 years to travel from Proxima b, to us on Earth. According to our astronaut's time the signal should reach Earth by February 2034. Now look at it from our time, 'Earth time'.

The planned landing of the rocket based on 'Earth time' was October 2030 (10.75 years travelling at 40 per-cent the speed of light from launch in January 2020). Add the 4.3 years it takes for the signal from the rocket to reach us on Earth, and you have a date of January 2035. The time that now separates us on Earth from our astronaut on Proxima b, is five years and four months made up of an eleven-month genuine time difference due to time dilation and four years four-months due to 'natural time delay'. Plus, if you add in the fact that the gravity (curving of space) is around 10 per cent stronger on Proxima b than it is on Earth, you get another, albeit fractional, variable in the rate of time (Headache anyone?).

There is no such thing as absolute time.

More disturbingly, perhaps, there is no such thing as 'now'. This second, for you, is different for everyone else. It's not that noticeable in everyday life, where the differences are measured in billionths of a second, but extend it to pilots flying 747s, to astronauts in the space station, astronauts on the Moon, and eventually astronauts on Mars and beyond, and the illusion of what we call 'now' gets exposed to the tune of seconds, minutes, months, years and thousands of years' difference. Because time is intrinsically linked to space. And space is malleable. It bends, twists and contracts. It's not consistent in shape. So, as a consequence, time stretches and contracts. It doesn't flow or tick at a consistent rate.

So, where are we? Let's summarise.

This thing with the totally inappropriate name 'space' is not nothing.

Space has an age. It's around 13.8 billion years old. Before that, the three-dimensional nature of space we experience and exist in, didn't exist – as far as we know. There was nothing for us to be born into.

Space, once started, has never stopped growing throughout that 13.8 billion years. Even more staggering, space is not just continuing to expand, it's actually accelerating. That is, it's getting faster, possibly moving outwards in all directions at speeds beyond the speed of light (only space itself can go faster than light).

Space is not empty.

Space is made up of 'quantum foam', a seething mass of impossibly small particles that flick in and out of reality. Where they go when they're not in 'our reality' is beyond me. At Planck length levels, space might be granular, made up of tiny, tiny, tiny quantum loops, more tiny than any tiny you can possibly imagine.

Space bends, twists, and does the 'quantum fandango' when it feels the presence of matter.

The more matter (or mass) present, the more space ties itself up in knots. Space is not just space. It's space and time fused into one single entity called space-time. When space gets warped, so does time. The more space is contorted, the slower time passes. At extremes, where space turns in on itself to create what we 'see' as a black hole, time slows to a virtual stop.

Space also feels and reacts to the presence of accelerated motion. It shrinks in the direction of this motion, causing time to slow in equal measure. Astronauts currently orbiting Earth in the space station are ageing 3.8 seconds a year slower than you and me, because of their acceleration through space. Find a way one day to accelerate a human being to millions of miles per hour, or to enable humans to live in extremely curved space and that 3.8 seconds would extend to days, months, years, even thousands of years, compared to us on Earth. And those 'long living folk' would have no idea time had passed any differently to us 'long gone folk' on Earth. To them, every day would just feel like a normal day. Because all the atoms that make up their 'biological clock' would tick slower. That is the weirdest thing of all.

So, on your next normal day, think on this.

When you're sitting in a chair at home, you are sitting 'in space'. But the nature of that bit of space you're sitting in, that surrounds you, is unique. It's space close to Earth and so it curves accordingly, because of the presence of Earth. If you were to pick your chair up and plonk it down at the top of Mount Everest, the space there would be different. It's less curved because it's further away from Earth. The difference in curvature at ground level and the top of Mount Everest is minuscule, but it exists.

And, when you're sitting on your chair at home, you're not just sitting in curved space, you are also sitting in time. While you're sitting, time doesn't just tick by, independent of the space you are in, it ticks in a way that's unique to that bit of space and so it is also unique to you.

Go back to the top of Mount Everest (you and your chair) and time will tick differently for you up there, as space will be slightly less curved. None of this is at any rate of difference you would notice, but it still happens. Plonk your chair anywhere near a black hole (just far enough away so as not to be shredded) and time would slow for you to a point where thousands of years would pass in the gently curved space on Earth, versus only a few hours for you in your extremely curved space.

That's just sitting (in your new, very exciting, chair). Get up and take a few paces forward and you're not just moving through space, you are moving through time. When you move through one (space) you also move through the other (time). And the weirdest thing of all? Even in the most extreme curved space, whether near a black hole or moving in a rocket at 67 million mph, you would think time is ticking along just fine. You wouldn't be aware that it had slowed to a virtual standstill in comparison to other people in less curved space. Because it's not just clocks that measure a slower time, it's your whole body clock. All the particles you are made of, your biology, if you like, react to curved space, so 'you' tick slower – the actual particles in your body. You're just not aware of it; you feel normal. (This, I have to admit, is 'boldly layman' at the limit of understanding just how this can be real. But it is. As far as every clever person in the world seems to confidently think. Plus, it's been scientifically tested and proven.)

So, that's space (or space-time). It took quite a lot of writing, to write about a 'nothing' that's supposedly 'empty'.

Let's move onto monsters.

CHAPTER 6

THERE'S BIG SCARY STUFF OUT THERE

We've established the universe is a gargantuan monster that is, in truth, of a size that's beyond our ability to truly grasp.

And, if that wasn't bad enough, the stuff that makes up the vast majority of the universe, the thing we call space, appears to have a life of its own. It's growing, multiplying, replicating and sprouting new bits of itself at a rate, on very large scales, that matches and potentially even exceeds the speed of light. It's a scary, big place, getting bigger by the second. So not surprising, I suppose, it's filled with some very scary, big things! Even in themselves, they are beyond comprehension.

Here's one of them: a star called VY Canis Majoris.

What's so strange about a single star? After all, we now know there's billions upon trillions of them out there. Well 'VY' isn't any normal old star. Let's start with something reasonably familiar, our star, the Sun. Our Sun, by any measure you like to think of, is seriously big. Compare it to Earth. The diameter of Earth (i.e. if you run a straight line through Earth's centre, from one side to the other) is 8,000 miles. Its circumference (all the way around the outside) is around 25,000 miles. And if you're a tiny, insignificant Earthling like us, Earth feels pretty big.

The diameter of the Sun, however, is 108 times bigger than Earth, some 864,938 miles (1,391,982 km). That's huge. Its circumference, 2,713,000 miles (4,366,150 km). That's even huger!

Look at those figures.

Earth's diameter   –   8,000 miles

Sun's diameter   –   865,000 miles (give or take a scorched mile)

Earth's circumference   –   25,000 miles

Sun's circumference   –   2,713,000 miles (that's two million, seven hundred and thirteen thousand miles, versus a puny 25,000 miles)

If these figures don't do it for you, let's look at it another way. You could fit 1,300,000 Earths inside the Sun. (That's our whole planet we're talking about!) One million, three hundred thousand Earths! This calculation is a bit tricky, as putting Earth as 'solid balls' inside the Sun as a 'bigger ball' would always lead to lots of bits of space between the Earth-size balls as they were crammed in together. The 1,300,000 figure assumes these spaces are filled up with bits of planet Earth. (You didn't really need to know that, but just in case some clever Dick challenges you on the number.) For a truly amazing visualisation of what 1.3 million Earths look like go to space-facts.com.[16] It is brilliantly done and mind-boggling.

I'm aware I've used the word 'monster' quite a bit already, but I'm sorry, the Sun is a veritable monster. A seething mass of nuclear fusion, operating at temperatures in excess of 15 million degrees Celsius and well over a million times bigger than us!

We really do have to be thankful it's 93 million miles away.

One last way of visualising the size disparity. If you flew around the circumference of Earth in a 747, at a speed of 550 mph, it would take you roughly 45 hours to complete a full circuit. That same journey around the Sun would take you a full seven months. The Sun is big. Gigantically big. So, what's the deal with VY Canis Majoris? Let's start with its diameter. We've got 8,000 miles for Earth, 865,000 miles for the Sun and for VY Canis Majoris?

It's 1,500,000,000 miles! You need to see this written out – one billion, five hundred million miles across! That's about 1,540 times bigger than our Sun!

Just look at this:

Earth's diameter	–	8,000 miles (feels pretty big from where I'm sitting)
Sun's diameter	–	865,000 miles (hugely, mega-big!)
VY Canis Majoris's diameter	–	1,500,000,000 miles (totally beyond belief!!)

Its circumference is even more fun. Remember Earth at 25,000 miles and the Sun at 2,713,000 miles. Well, VY's circumference is 1,906,000,000 miles!

Earth's circumference	–	25,000 miles
Sun's circumference	–	2,713,000 miles
VY Canis Majoris's circumference	–	1,906,000,000 miles (close on 2 billion!)

Now, here's the really scary bit. We think our Sun is big, because we can fit 1,300,000 Earths inside it, but you only really get a sense of what big is really like when you learn how many of our 'monster' Suns could fit inside VY Canis Majoris.

The answer is ridiculous. Incomprehensible. Impossible, even. It's 9.3 billion!! You could fit nine billion, three hundred thousand of our Suns inside the gargantuan monster that is VY Canis Majoris. How anything that big can exist is quite beyond me. But it's out there. A huge, bloated monster of all monsters, floating in deep space (watching, waiting, like all monsters do).

There are some things I'm not sure I want to get a glimpse of. VY Canis Majoris is one of them. I don't have an irrational fear of big objects (megalophobia), but I have a feeling too much imagining and 'VY' could bring this particular condition on. If 9.3 billion Suns isn't a bowel-loosening figure, then what about us? How many Earths could you fit inside VY Canis Majoris? You won't like it. I don't like it.

You could fit around 70,000,000,000,000,000 Earths inside VY Canis Majoris. That's 70 quadrillion Earths! (Chapter 1 'Big Numbers'; don't tell me you've forgotten already!) If a reminder helps, a quadrillion is a thousand trillion, so that's seventy thousand trillion Earths, if you can possibly get your head anywhere near that number. VY would just gobble them up.

What more can one say about it?

Well, let's take our 747 flying at 550 mph. (Just the thought of boarding and I can feel my newly found megalophobia rising!) To fly around the Earth at this speed, done in 45 hours. To fly around the Sun at this speed, seven months. To fly around VY Canis Majoris at this speed? That will be... 11,000 years. (Sorry, that has to be a third... GULP!!!)

You'll need to get time off work. But me, personally, I wouldn't go near it with an intergalactic bargepole. To fly around the circumference of VY in a 747 you would have to have set off just as Stone Age man was building what is believed to be the world's first human-made site of worship, the Gobekli Tepe, in Southeastern Turkey, some 9,000 years before the Pyramids of Giza were built and 6,000 years before Stonehenge. And you would be arriving just about... now.

You might be wondering at this point where VY is, exactly, as it's not something you would want to bump into unexpectedly. Well, we're just about OK for now. VY Canis Majoris is around 5,000 light years from Earth and it's at times like this you really appreciate that a light year equates to 5.9 trillion miles – 5.9 trillion miles times by 5,000 is the kind of distance I want this thing to be. Because of its brutal size, VY Canis Majoris is appropriately called a 'red hypergiant star' and the person whose eyes must have popped out on stalks when he first saw it was a French astronomer called Joseph Jérôme Lefrançois de Lalande. That was way back in 1801. Can you imagine coming across the reality of this monster (VY now having redefined the word 'monster') all the way back in 1801? And yes, it's still there, 'waiting'. But, in cosmic timescales, not for long. It's anticipated VY Canis Majoris will blow up in what's called a hypernova in around 100,000 years' time. That will be some 'blow up'. Are there any other nightmare stars like VY lurking in the depths of space? Yes, millions, billions and trillions of them, such is the scale of the universe. And it's more than likely some will be even bigger. We just haven't found them yet.

How big might these others be?

Well, VY Canis Majoris is around 1,540 times bigger than our Sun. Current estimates of the largest possible star we might find reach up to 2,600 times the size of our Sun. That would mean not far off twice as big as VY Canis Majoris.

Oh, dear.

The next time planet Earth's daily dose of frustrations start to wind your coil – the late-again train, the ungrateful boss, the child that just refuses to do what it's told – conjure up a mental picture of the impossible enormity of VY Canis Majoris, all 9.3 billion Suns' worth of it. The reality of its existence, and trillions more monsters like it, might help put life into perspective a little. Well, it does for me.

Now, VY is just a star, comparable to our star, the Sun. What about our galaxy, the Milky Way... does it have any comparable monsters? Well, yes. This little baby – IC 1011.

What IC 1011 misses out on in the name stakes, it makes up for in sheer brutal size. IC is estimated to be in the region of 6 million light years across and, as such, holds the current record for the largest known galaxy in the observable universe. Think about the enormity of our own Milky Way, at 100,000 light years in diameter. It would take us, at our current 'fastest speed ever', around 2.8 billion years to cross the Milky Way. IC 1011 is 60 times bigger! (Don't bother with 2.8 billion x 60. I mean, really, don't bother.) Even light travelling at the humongous speed of 5.9 trillion miles a year is going to take a full six million years just to get from one side of IC 1011 to the other! Six million years!!! (It's light! It's the fastest thing we have!)

And stars? We think our Milky Way has an unfathomable number of stars, some 400 billion. Well, just imagine what this behemoth of a galaxy IC 1011 has… around 100 trillion stars. One hundred trillion stars, just in this one galaxy! It's truly mind-blowing. If you were to put IC 1011 in place of our Milky Way, it would gobble up the Milky Way and its 400 billion stars. Continue to gobble up all of the 2.5 million light years of space to the Andromeda Galaxy. Continue to consume the mere 220,000-light-years-across Andromeda and its trillion stars. Onwards for another 500,000 light years of space until it finally gobbled up a galaxy called the Triangulum Galaxy and its 60,000-light-year diameter and 40 billion stars. That's some voracious appetite. And truly, some kind of monster.

For the record, IC 1011 is called a supergiant (very appropriate) elliptical galaxy and it lies around one billion light years from Earth (good). Now, what's amazing about this distance, one billion light years, is IC 1011 was first discovered on 19 June 1790, (that's 1790, not 1970!) by a British astronomer with the very un-British name of Frederick William Herschel. On finding it, Herschel described it as a 'nebulous feature', not recognising it as a galaxy in its own right. And, of course, not recognising he had identified what is still the largest-known galaxy in the observable universe –

over 300 hundred years ago! You have to hand it to Herschel. If VY Canis Majoris and IC 1011 are a megalophobic's worst nightmare, then the next little monster nestling within the vast regions of space is going to give claustrophobics a very sleepless night. It's the neutron star, the spawn of a giant star's death (check under the bed, look in the cupboards, buy a night light). Even the word 'neutron' has a compressed feel to it, it's not 'feather star' or 'fluffy star'. And, as we'll see, these deeply dark little fellas do small, compressed and heavy equal to the extremes that stars like VY do big. But before we expose ourselves to the mind-bending reality of the neutron star, a little step back to stars, how they are born, live and die (it will help).

Now, clearly this stuff is complicated – the birth, life and death of stars. It's books' and books' worth of writing, but not for this book. Because this book is not meant to be complicated, it's for us normal folk. So, for those of you out there who really know your stuff, apologies, but I'm going to be cutting quite a few cosmic corners.

In space, there are gas clouds; vast oceans of the things that have been floating around for billions of years. And they can stretch across hundreds of light years, so they're big. And these clouds are full of stuff. Stuff like molecular hydrogen, helium, heavier elements and, thankfully something I do understand, dust. In such huge clouds, this stuff or 'matter' (the more technically correct term) can add up to millions of times more than the matter or stuff in our Sun. And remember, relative to Earth, our Sun is huge. So there's vast amounts of stuff, and what all this stuff wants to do is clump together and collapse the cloud under the force of gravity (space curving and bending downwards due to the presence of matter).

Against this somewhat suicidal tendency of matter is the gas cloud's inherent pressure, which attempts to keep the cloud inflated, or 'up'. And these clouds live like this for millions, to hundreds of millions, to billions, of years. It's a delicate balance of push and pull, the pressure just overcoming the pull of gravity to keep the cloud inflated.

A simple, if not very exact, analogy. There's a lilo on the floor in front of you. Because there's no pressure inside (air) the lilo material (matter) has collapsed under the force of Earth's gravity. You start blowing it up, puffing and puffing until it's fully inflated. Then, bugger me, just as you are going to jam the stopper into the air valve, one of your friends, all 14 stone of him, plonks himself down onto the lilo. His body weight (mass) adds to that of the lilo, and gravity becomes dominant again, squeezing the air out of the lilo as it flattens out. You dive for the air valve in a frenzied state of red-cheeked puffing, trying to create enough new air pressure to counterbalance the weight of your friend and the lilo material.

Let's assume you've got the lungs of an African elephant and, somehow, you start pushing the lilo – and your inconsiderate friend – back up (which would mean you're puffing marvellous!).

But then, lo and behold, a second friend joins the first, adding another nine stone of body weight onto the lilo. No lungs on Earth are going to counteract that. The pressure you are trying to create by blowing air into the lilo (air pressure) is totally overwhelmed by the combined mass and weight of the lilo material itself (which, of course, is tiny) and your two friends (who are the real problem). Gravity overwhelms the air pressure and the lilo deflates.

This is what eventually happens to large gas clouds. The matter inside the cloud increases, maybe through merging with another cloud, and as a result, the density increases, causing space to curve and twist downwards, driving the cloud to collapse in on itself. Just like your lilo. Down and down the cloud goes (hundreds of light years of it!), the matter within it being compressed into a smaller and smaller place under the force of gravity. As the matter is compressed, it creates more and more energy and so gets hotter. And we are talking off the scale hot. This collapse continues until the temperature of the compressed matter gets so extreme that nuclear fusion erupts (think nuclear bomb then add a few trillion).

This 'nuclear event' creates a huge burning core at the heart of the collapsed cloud (which doesn't really resemble a cloud by this stage) and this burning core causes the collapse to stop, as it pushes energy outwards, creating a new kind of pressure. This pressure pushes all the remaining matter that's not part of the burning core back out until it all settles into a new kind of equilibrium. The matter outside the core still wants to collapse inwards, but it is held at bay by the pressure (heat energy) generated by the burning core. And, hey presto, we have a star! (You can think of it as getting your lilo back in a slightly different shape. It also happens to have heated up to around five trillion degrees centigrade.)

And so, to coin a phrase, 'a star is born'. Many are like our Sun. Others, like VY, are ludicrously bigger (9.3 billion times bigger!). And there they hang, the push and pull of nuclear fusion and gravity balancing each other out, for millions upon millions and billions of years. It's a wonderful thing.

Another wonderful thing? It's estimated around 480 million stars are born every day in the universe (480 million!). That's around 5,500 new stars born... every second! Put that in your cosmic pipe and smoke it. The next time you're floating on a lilo in the Med, spare a thought.

Fast forward 10 billion years. Gravity (curved space) which has been held at bay all this time, desperate to collapse the star in on itself, finally senses its time has come. Because the outward pressure of the nuclear fusion going on inside the core is waning. Put simply, it's running out of fuel. For a staggering ten billion years or so our star, which was a cloud, has been burning up millions of tonnes of hydrogen every second, creating the outward energy needed to keep it afloat. But, when the hydrogen runs out, big stars (ones that are more than six times the mass of our Sun) don't just give up. They turn to burning helium, then carbon, then oxygen. When that's all gone, they turn to stuff most of us probably didn't even realise could burn – silicon, aluminium, potassium. Like your worst kind of pyromaniac, these stars burn everything they can get their hands on to keep themselves inflated. But, eventually (after a few million years), they meet their match.

Iron.

Now, even I had a sense iron would do for them. And it does, because no energy can be created, not even in the raging infernos of a star in its death throes, by fusing together iron atoms. Our star is done for. At least, in its 'shiny' form. What happens next, no words can really do justice to, such is the scale and ferocity of it. We're talking total cataclysmic collapse. Not the kind of collapse you see on the news, when an old tower block is brought crashing down by neatly placed explosives. This is the collapse of a whole star. A star whose diameter is millions, perhaps billions, of miles across, that you could maybe fit 70 quadrillion of our Earths into, collapsing in a fraction of a second into an entity just 12 miles wide, give or take a mile. From billions of miles across to just 12, faster than you can say 'one', with the collapsing matter reaching speeds in excess of three million miles per hour. That's gravity for you, or curved space, the thing that keeps our little feet planted gently on the surface of planet Earth. But here, with 'our star', it's gravity manifest as the most destructive force we know of in the universe. And what gravity creates from the death throes of a large collapsed star is the 'devil star' – the neutron star.

You'll remember as our gas cloud collapsed that matter got increasingly compressed into a smaller and smaller space until it reached such an extreme temperature it erupted into a burning core, to create our original star. Not here. There's nothing left to burn. The star burnt up everything. It's dead. So, all the in-falling matter of our star just continues to get crushed into an ever smaller and smaller space, a cosmic squeeze of such gigantic proportions that not even the atoms that make up visible matter can withstand the forces involved. They wilt, bend and eventually melt, as they are pummelled out of existence. This is atoms we're talking about. Those tiny, infinitesimally small things that make up the air we breathe, you, me, overpaid footballers and buses.

All crushed to oblivion.

And worse. Even the tiny particles inside the atoms, the electrons and protons, can't withstand the pressure. They dissolve. Nearly everything that makes up existence is gone, crushed, liquefied, mashed, until there's just one thing left standing.

You know what it is.

The neutron particle. Trillions upon trillions of them, squeezed impossibly tightly into a dark, menacing ball. So repugnantly dense is this dark black heart that in the instant it forms (and we're talking fractions of a second) it acts as an impenetrable barrier to all the other in-falling matter. This matter, still cascading down at speeds of over three million miles per hour, hits the newly formed neutron ball and rebounds in a gigantic shockwave, creating one of the most powerful energy sources in the whole universe – a supernova. How powerful? Think about how bright our Sun is, even though it's 93 billion miles away, and times that by around ten billion. It's so bright that for a brief moment a supernova can outshine all the other stars, combined, in its own galaxy. And, when the brightness of the supernova has subsided and the dust settles, we're left with a dark shadow nestled in the deep blackness of space.

The neutron star.

Where do you start with this malcontent, this cosmic abomination? Tell it as it is, I think. Because so mind-boggling are the facts, they can't be embellished. First, its atmosphere. All objects in space have an atmosphere, don't they? In terms of the neutron star, the answer is 'just' or 'hardly'. While Earth's atmosphere extends to around 62 miles above its surface, the atmosphere above this manic ball of condensed neutrons is just – four inches! Just four inches above the surface!! It would almost be laughable if the neutron star weren't so menacingly, deadly serious. Even more brain-cell-melting, while our 62 miles' worth of atmosphere is full of lovely, light, life-giving air, the four inches of atmosphere above a neutron star has a density similar to that of a diamond! Ever tried breathing solid diamond? It's truly hard to grasp; and that's just the atmosphere, we haven't even got to the actual surface yet. You know that is going to be something else.

Think of steel, the stuff we build our skyscrapers with. It's what trains run on (the tracks). It holds up bridges across huge rivers and canyons. You get steel. So if I was to come to you one day and say, 'Hey, I've found a metal twice as strong as steel,' you'd likely go, 'Wow, that's pretty impressive, twice as strong as steel.' If I was then to say, 'No, actually, it's a hundred times stronger,' you'd probably go, 'Wow, a hundred. I mean, how can anything be a hundred times stronger than steel?' And then I'd go, 'Well, actually, it's not a hundred, it's a million... well, no, it's actually a billion... well, it's not even that, the metal I've found is actually... ten billion times stronger than steel.' You wouldn't get it. Neither do I.

Just like the distances involved in space, especially the infinite one, we have no mental reference point for realities such as this. How can we possibly stretch our imagination to conceive of a substance not just twice as dense and strong as steel, but ten billion times as dense and strong? But it's out there. It's real. And it's just the crust that forms the surface of a neutron star.

What is the crust like to look at? Quite eerily, it's glassy smooth. All that sucking gravity pulls every neutron in as tight as nature will allow, creating a surface that resembles a polished ball bearing. Whatever imperfections there are on the surface only rise to a massive, wait for it... five millimetres. That's a neutron star mountain for you, all five millimetres of it, versus our Mount Everest, at 27,000 feet.

What kind of gravity does this?

Think of Earth and the gravity that keeps us from floating up into space. The gravity on the surface of a neutron star is a stomach-clenching 200 billion times greater! Jump off a wall 20-feet high onto a hard, concrete pavement. You would feel the force of Earth's gravity as you were falling through the air, even more so as you hit the concrete surface (both legs or more, busted to bits). Now imagine, or at least give a moment's thought to, a gravity 200 billion times stronger than that.

So, that's the outside. A four-inch-thick diamond atmosphere with a surface crust 10 billion times stronger than steel, exerting a gravitational force 200 billion times greater than that on Earth.

Dare we go inside?

The first thing to say about the inside of a neutron star is, it's beyond any kind of weird you will ever have come across. Below the crust, our 'billions-of-times-smaller-than-a-pinhead' neutrons are compressed to such an insane density that conditions reach upwards of one billion degrees Celsius! (The Sun, by comparison, at its centre, is a mere 15 million degrees Celsius. Almost on the chilly side.) And what this unrelenting compression and resultant heat creates, by most physicists' reckoning, is what they call a 'superfluid' – a kind of plasma. Now, at first, I didn't find the idea of a superfluid particularly dramatic. To be honest, I was slightly disappointed. I thought the dark star would deliver up more. That's until I read that scientists have managed to create a watered-down version of this superfluid in a lab environment here on Earth. Now, as you might expect, they contain this stuff in very well-sealed, airtight containers. But, in one experiment, it appears this superfluid replica actually climbed up the inside of this 'very well-sealed airtight container' and escaped out of it... all on its own! Now superfluid has definitely got my undivided attention.

Thankfully, the superfluid we're able to create on Earth is like a real 'baby version', much more lightweight, not the monstrously thick, horrendously fat superfluid found inside the neutron star. (But we've still made some, and it's still here on Earth. Where is that escaped one, I wonder?) So, how to get a handle on the true nature of neutron star superfluid? Well, take a sugar cube. It would be good not just to picture it, but to get one in front of you. Now, imagine, if you can, all of you, your whole body, crushed to fit inside this little sugar cube. That's a lot of body (depending on your size) squeezed into a very, very small space, about half an inch (1.2 cm) cubed.

Because your body has been forced to leave the relatively comfortable space it occupied in your skin and has been compressed into a volume in the region of 80,000 times smaller, it becomes denser (compressed, squeezed more tightly). Not only is the sugar cube now denser, because it's got all of you in it, it also weighs more, because it's got all your weight in it – or mass, as we now understand it, from Chapter 4.

Now, take this a little further. Find a friend, someone you feel close to, and imagine he or she has now joined you inside the sugar cube. Two whole bodies now forced to fill the space your crushed body was just getting comfortable in. The mass inside this little sugar cube is now even more dense. Think of any unsuspecting coffee or tea-drinker trying to pick it up, now it's got you and your friend's combined mass and weight in it. They couldn't, and yet it's just a little half an inch cubed piece of sugar.

You get the principle. Density is a measure of how much material – 'matter' – has become compacted (in this instance you and your unfortunate friend) into a given volume of space (in this instance the sugar cube). And we've also understood that material or matter has weight. And, in this example, the combined weight of you and your friend is transferred into the volume of space you have been crushed into – a volume the size of a sugar cube.

So, we have a very heavy sugar cube. How dense could it get, and as a consequence how heavy could our little sugar cube get, so we get close to the density of neutron star superfluid?

OK, let's get you and your friend some company. How about the New Zealand All Blacks?

They've got a pretty impressive combined mass and weight and they're used to sharing a bath together, so let's put the whole team in. That's a 31-man squad with an average weight of, say, 17 stone. Now, not only do we have a very cramped and dense sugar cube, but it has a weight of 550 stone – including you and your flattened friend. (Try to picture that little sugar cube with all that dense body mass crushed into it, carrying all that weight.) But let's not stop there, when there are so many other rugby players at our disposal. Let's gather up all the big old rugby players that took part in the 2015 World Cup in England. And why not? That's another 19 teams, 31 players in each squad – 589 players. Crush them all into your tiny sugar cube with your autograph-hunting friend.

We now have an incredibly dense sugar cube; with each player averaging 17 stone, our little cube of sugar weighs close on 10,563 stone. That's not far off 68 metric tonnes. Dense enough? Heavy enough? Not quite. Let's go mad and include all the fans, officials and volunteers that attended the World Cup quarter and semi-finals – 460,732 – and the final at Twickenham – 82,000. That's you, your ever-grateful friend, and over half a million people, all now sugar cube size pulp! (No elbow room at all, or to be more accurate, no actual elbows!) Try to imagine, if you possibly can, the density of that little cube. All those bodies, roughly 543,000 squeezed into that tiny volume of space, just a half inch cubed.

Now think of the weight that little sugar cube carries, because even though 543,000 people have been reduced to a half inch cubed space, their weight remains. And all their weight (remember you're in there... somewhere!) is transferred to our little sugar cube. I'll do the maths, so you don't have to – 543,000 people based on an average weight of 12 stone per person. (The average weight of a man in England is 13 stone and a woman 11 stone, so call it 12 stone as an average.) This gives you 543,000 people x 12 stone average. The grand total? A mighty 6,516,000 stone (six and a half million!). All in your sugar cube.

How heavy is that?

Well, no crane in the world could lift your sugar cube, such is its enormous weight. There are big, big cranes in the world, gigantic mobile cranes. Monster cranes, which are, in fact, vessels at sea. And the most humongous in terms of lifting power are gantry cranes. And the daddy of all gantry cranes is the Taisun gantry crane, housed in Shandong Province, in China. This little baby can lift close on 22,000 tonnes! And it currently holds the official Guinness World Record for lifting. But – and no one thought about this when awarding this mother of all cranes such a prestigious award – it ain't gonna lift our sugar cube! No way!

Because our little sugar cube, weighing in at over six million stone, equates to around 41,000 tonnes! The Taisun gantry crane is well and truly done for. So there you have it. Look at your tiny sugar cube, if it's in front of you. There's you, your friend, more international rugby players than you can shake a stick at and around half a million fans, with a combined weight of over six million stone, which equates to around 41,000 tonnes. And no crane in the whole wide world, however much it tried, could lift this single sugar cube!

So, I hear you say, what's this got to do with the superfluid inside a neutron star? Do we even remember superfluid, such has been the detour into sugar cube land? Well, it's got a lot to do with it. Because you need an analogy like this to have any chance of understanding the quite horrific density and weight of this superfluid. So here it is then, the leap. Take the population of London, 8.6 million. Times it by an average of 9.5 stone (average global weight). You get 518,000 tonnes. All of them, with you, in your sugar cube. But that's not superfluid power.

Take the population of the UK – 65 million – and times it by 9.5 stone. You get 3.9 million tonnes! It's not this, either.

Take Europe, sauerkraut, onions, pasta and all. That's 743 million people. Times it by 9.5 stone and you get 44 million tonnes. You'd think that would be enough, wouldn't you? It's just a sugar cube, for heaven's sake. (By the way, how are you feeling in there... not so neighbourly, after Brexit?) No, to match the density and the weight of the superfluid you find inside the core of a neutron star, you have to crush the entire human race down into your sugar-cube-sized space. That's all 7.6 billion of them. (I say 'them' because I'm personally opting out of this little exercise.) So, it's 7.6 billion people at an average of 9.5 stone. You get a sugar cube weighing in at around 458 million tonnes!

Our (or should I say 'your') sugar cube now weighs the combined weight of the entire human race (less me). That's 458 million tonnes. You would think that was it, wouldn't you? But this is a neutron star we're talking about. A poisonous, solid little dark dwarf of a thing. And this is its heart we are describing. The stuff that, when on Earth, climbs out of an airtight container of its own accord! No, this superfluid insanity, just a sugar cube's worth of it, would weigh three times the total human race – 1.3 billion tonnes! (That needs more exclamation marks – !!!!!!! and !)

What does a 1.3 billion tonne sugar cube feel like?

Well, take your sugar cube-sized little lump of neutron star superfluid and suspend it one mile above the Earth (no, don't ask me how) then drop it. This little baby, all 1.3 billion tonnes of it crammed inside a sugar cube, will fall for a mile, averaging a speed of around 700 million miles an hour, and go straight through planet Earth and out the other side, probably taking a few kangaroos with it for good measure. Such is the sugar cube's compressed mass and weight, it will go through Earth (that's around 8000 miles of rock, lava, solid iron, your house) before you can say, 'What the...?'. It will then continue on its way, probably slicing through other planets, comets, stars and everything in its path like a hot knife through butter. Probably into infinity and beyond.

What's really unbelievable is, this stuff is real. It's not out of some pointy-eared Trekkie flick on the Sci-Fi channel. It's out there, my friends! And the very point of this book is to attempt to make us sit up and pay attention to the stunning reality of the world, the universe, the reality of what you (yes, you!) are part of. So the next time you're head down on your mobile, causing people to shuffle frustratingly behind you, or you're sitting in your favourite coffee shop drinking your venti soy quadruple shot latte, extra hot, no foam: take a moment and look up. Think about that 1.3 billion tonne sugar-cube-sized bit of neutron star superfluid carving its way through planet Earth quicker than you can say 'vindaloo'.

One other thing about neutron stars, as if there already hasn't been enough. They spin. But you know now this isn't going to be spinning in any normal sense of the word. Let's take the huge cosmic rock we are currently standing on, planet Earth. We don't feel it, of course, but our Earth makes one complete rotation every 24 hours. (It's why we see the Sun rise, see it set and see it rise again 24 hours later. The Earth spins, albeit in a quiet leisurely way.)

The speed any one of us is travelling at, as the Earth spins on its axis, depends on where we are on the planet. This can be 0 mph if you're right in the centre of the north or south pole or up to 1,045 mph/1681 kph if you're standing at the equator. (If you live in London, for instance, it's around 600 mph/965 kph. It's amazing we have no sense of it.)

So, one spin or rotation every 24 hours for us Earthlings, ranging from 0 mph to 1,045 mph (pretty fast). But what if you were a neutron star-ling? Well, just like everything else about this angsty little ball of beads, it's enough to make your head spin. There's a neutron star that goes by the snappy little name of PSR J1748-2446ad (don't marry

an astrophysicist, think what your kids will be called!).
To ease the name strain, let's do what all parents hate you
doing and shorten it to PSR.

Are you ready for this?

PSR spins 716 times every second!! On Earth, we're talking
one spin every 24 hours; this particular neutron star does
716 spins in a second. (Say 'one hundred' – PSR just span a
full rotation 716 times!) How can anything 12 miles wide
and so buttock-clenchingly dense spin that fast! What's that
a minute? An eye-watering, contact-lens-spinning 43,000
rotations! What is it in just one hour? It's 2,577,600
rotations. (I'm sorry, but I have to spell this out – over two
and a half million rotations in just 60 minutes.) And, finally,
the denouement! While we spin just once in 24 hours, this
manic little 12-mile-wide ball of superfluid, nicknamed
PSR, with a crust 10 billion times stronger than steel and an
atmosphere as solid as diamond, spins... sixty-one million,
eight hundred and sixty-two thousand times. (I need a new
book just to fill it up with exclamation marks!)

Try, if you possibly can, to imagine all that density and
weight of superfluid (remember, just a sugar cube's worth
weighs 1.3 billion tonnes), spinning over 61 million times
between you waking up, going to bed, and waking up again.
And yes, as I've said and will continue to say, it's out there!
A monster cosmic spinning top that's spinning close to a
thousand times with every breath you take. A recap, I think.
You don't have to, but I find it helpful.

So, we have mega-big, light-years-wide gas clouds. They float around for millions into billions of years, the gas pressure just enough to keep all the matter from caving in under the force of gravity. Eventually, as it always does, gravity (the curving of space) wins.

The cloud collapses, all the in-falling matter gets more and more compressed, which creates more and more heat, until it erupts like a North Korean fireworks party. The thermonuclear eruption creates a new pressure that halts the collapse, and a star is born. We hang around for a few more billion years until the burning nuclear core burns itself to a cinder. Without the outward pressure the star collapses (gravity at it again).

Now, there's a point I didn't tell you about here. If the collapsing star is less than nine times the mass of our Sun, then all the melting atom stuff I mentioned previously doesn't happen. The star collapses to another star form, called, of all things, a white dwarf. It's a name that feels almost 'cuddly pet like' compared to the bonkers ball neutron star. (The white dwarf is still pretty potent in the density and weight stakes, but not on the same planet as the head-bashing neutron.) When a collapsing star is over nine times the mass of our Sun, that's when the in-falling matter completely ignores the cuddly pet stage and plummets to neutron star extremes. What eventually stops the collapse is a law of physics that doesn't allow neutron particles to occupy the same space, so they just cram together as close as they can in a cosmic game of sardines.

As the core of the neutron star is formed, it creates an impenetrable barrier that all the other in-falling matter rebounds off, all within a fraction of a second. And you get a supernova, one of the biggest known explosions in the whole of a humongously large universe. An event that is so 'words-can't-quite-describe-cataclysmic' that for a brief moment it outshines the brightness of all the other stars in its galaxy combined (we're talking over 100 billion stars). Such is the force of this explosion that all of the collapsing star's remaining matter is propelled thousands of millions of light years out into space.

This 'dead star stuff', by the way, eventually lands on a funny little piece of rock circling another star which is still in its 'inflated stage' (i.e. 'on' or 'alive'). The 'dead star stuff' interacts with other funny chemical stuff on the small piece of rock and eventually it creates you! (Yes, you!) That's what we are. The result of a collapsing star exploding outwards. Star stuff that travelled for aeons, across billions of light years, landing on the rock we call Earth. A bit of chemical interaction over a few billion years, simmered gently by the heat of our star, the Sun, and it's another 'hey, presto' – human life. We are literally the stuff of stars. A small miracle, by any stretch of the imagination. And while all this palaver is going on – supernovas and the creation of human life, etc. – the ugly duckling of the star world has formed: the neutron star.

What was once a bright, heat-giving, life-giving force is now an impenetrable ball of black neutrons squeezing the life out of each other. Dark. Dense. Deadly.

Should you be worried at night as you're pulling the bedcover up to your chin?

One hundred million.

That's how many neutron stars are estimated to be in just our galaxy, the Milky Way. One hundred million! (Pass me the Mogadon.)

So, there's human-shredding monsters out there.

Spinning monsters, like the neutron star. Mega-bloated, impossibly big monsters, like VY Canis Majoris. And inconceivably vast behemoth structures, like the yawning IC 1101, all six million light years of it.

But while these are truly knee-trembling monsters, at least they are 'there'. You might not want to see them coming, but if you need to, you can. But what if there was nothing? Just total, endless blackness. Isn't that the ultimate fear?

Welcome to 'The Void'.

An incomprehensible, 1.8-billion-light-years-across stretch of space. And 1.8 billion light years' distance of total blackness. No galaxies. No stars. No planets or moons. Nothing. Just a black, cold, forbidding and seemingly endless cosmic hole where 10,000 galaxies should be! And we are told it shouldn't exist (which is a little worrying).

'The Void', we are told, defies all theories of large-scale structure formation in the universe. And it even challenges advanced computer simulations that struggle to replicate it. And yet it's there, in all its huge 'nothingness'. (Other than space-time itself.) A chap called István Szapudi, at the University of Hawaii at Manoa, has crowned it 'the largest individual structure ever identified by humanity'.

Me? I'd crown it... 'Aghhhhh!!'

Where is it? Well, you can let out that breath for the time being. It's between 6 to 10 billion light years away (phew!).

Let's hope it doesn't grow much bigger any day soon.

THE BOGEYMAN

So that was the humongous VY. The neutron star – a deadly dense ball of spinning madness. The monstrous galaxy-gobbling IC 1101 and… 'The Void'. What on Earth (or rather what in the universe) could possibly follow these horrors?

You know what's coming.

This is the tarantula at the bottom of your bed. The infestation of bats crammed into your crawl space. This is the nightmare you force yourself to wake from… only to find the bogeyman is real and standing in the shadows in the corner of your room. This is the hole. The never-ending black hole that nests itself in the dark recesses of space. A monstrosity of horrendous proportions that tears the very fabric of space-time and shreds and destroys everything the universe has taken 13.8 billion years to create. Let's get you there gradually, to understand first, how they form.

Imagine three blocks of flats that are due for demolition. Let's say each one is twenty storeys tall. If the explosive experts have done their jobs correctly all three blocks of flats will collapse completely, but they don't. Block 1 collapses down to floor 10 and stops. The top 10 floors (albeit a bit wobbly) are still upright above ground. Block 2 collapses further, to floor 15, and stops. The top 5 floors are still upright above ground.

Block 3 does as the explosives experts planned and it collapses to the ground. You see nothing upright. Nothing that resembles the original Block 3 of the flats. It's just dust and rubble. This is similar to the three possible phases large stars can go through when the fuel they've been burning at their core, for billions of years, eventually runs out and their mass (matter) starts to collapse inwards under the force of gravity.

If it's a star one and a half times bigger than the size of our Sun, it will do what the first block of flats did: collapse some way, then stop, as a new kind of pressure kicks in. In Block 1 it was the 10th floor that decided to stop the collapse. In our star, what stops the collapse is something called 'electron degeneracy pressure'. The energy created by the gravitational collapse isn't enough to crush or melt the tiny electrons inside atoms. So the electrons bunch tightly together and act like a 10th floor. They stop the collapse, which creates the pressure (think base or foundation) that pushes the remaining in-falling matter back out. The star version of the 10th floor is the white dwarf. As we mentioned in the last chapter, a white dwarf still possesses a mean amount of density and weight. Ship a bit of white dwarf back to Earth and put it in our sugar cube and the sugar cube would weigh around five tonnes. That's a tiny sugar cube, weighing the same as an African elephant. And, just like in Block 1, where we can still see something resembling a block of flats (albeit just the top 10 floors), so in the white dwarf we can still see something resembling a star. And, because it still carries some of the residual heat of the original star, it will also 'shine like a star' for a while (a few million years, give or take).

To Block 2. Well, stumpy little Block 2, just 5 floors sitting atop a pile of rubble, is the block of flats version of our malevolent little neutron star. It happens when a star one and a half to three times greater than our Sun collapses. The greater amount of matter, or 'stuff', in these bigger stars creates a greater gravitational energy (remember, the more matter there is, the more space-time warps and curves downwards).

In this case, the energy created from the in-falling and increasingly compressed matter crushes the tiny electrons that were holding up the white dwarf to smithereens (floor 10 just got obliterated). The collapse continues, with increasing energy, until it reaches a point where just the neutron particles inside atoms are left. Even though these bigger stars have collapsed through the white dwarf stage, liquidising the electrons, they don't have enough gravitational clout to crush the neutrons. The neutrons bunch together and form their own base or platform that can't be crushed any further.

Just as we had 'electron degeneracy pressure' holding up the white dwarf, we now have 'neutron degeneracy pressure' holding up the neutron star. And, just like Block 2, where we can still see something just about resembling a block of flats (5 floors on a large pile of rubble), so we have something vaguely resembling a star. (I say 'vaguely' because it doesn't burn, it doesn't shine and you can't see it – lucky to hold onto the 'star' label, methinks.)

So, to Block 3 (no one wants to live in Block 3!).

Nothing could stop the collapse of this tower block. It pulverised the 10th floor, smashed through the 15th floor and left nothing remaining that bears any resemblance to the original block of flats. In fact, even the part of this analogy that talks about 'piles of rubble' is misleading. It would be more accurate to talk about a collapse that doesn't just take down the whole of Block 3, but continues, driving down into the very earth itself, causing a horrendously large, gaping hole to open up in the ground where Block 3 used to be. A hole that not only appears to be swallowing all of the debris and rubble from Block 3, but a hole that's actually growing, sucking in the debris and what remained standing of Blocks 1 and 2.

If you happened to summon up the courage to walk to the edge of this hole (steady!) and stare down into the ever-widening abyss, you would be looking into a basophobic's worst nightmare (that will be the fear of falling). A hole that goes down forever. Without end.

You could drop a stone down this hole and wait until the end of the universe, and still you wouldn't hear the 'plop'. This stomach-churning event is what happens when stars 25 times the mass of our Sun implode. The monstrous gravitational energy, generated by the star's imploding mass, pulverises everything. It rips atoms apart, spitting out their electrons and splitting the nucleus, laying bare the protons and neutrons. The relentless energy then melts the electrons and fuses together the protons and neutrons to create just neutrons.

And then, even these unbelievably solid and deeply dense particles, the ones that make up the seemingly impregnable neutron star (just the crust 10 billion times stronger than steel!) experience Armageddon, as they, too, are liquidised, to leave only the quarks standing. What follows next is truly astonishing.

Existence itself, as we know and understand it, disappears.

The catastrophic scale of the collapse and the unimaginable scale of the energies involved act like a cosmic meat grinder, shredding even the quarks (not the quarks!!!!). Every known particle that makes up the living world, ground to a pulp. Everything that we know of or can describe, even space-time itself, gone.

Think about this for a moment.

We are talking about stars like VY Canis Majoris, with a diameter of over one and a half billion miles (repeat that slowly), which you could fit over nine billion Suns into, imploding. In just fractions of a second passing through the white dwarf stage, (that's a one-and-a-half-billion-mile diameter down to around a 13,000-mile diameter in less than a blink of an eye!) down to the size of a 12-mile-wide neutron star. But not stopping there. Imploding to six miles wide, three miles, one, then to the size of a football, an orange, a pea. Down and down, with no known particle in the universe able to halt the collapse. Down to the size of this full stop [.] if you can possibly get your head into this full stop. The whole of what was the core of the VY Canis Majoris... here [.]

And yet, still nothing to stop it. Nothing to stop the relentless 'dive'. Down to the size of an atom, something so small you could fit five million of them in the tiny full stop in the brackets above. Still down and further plunging down, the gargantuan, nightmarish power of gravity sucking everything into a smaller, tinier, more infinitesimal space until... you reach a point of zero volume.

A singularity. (This is not a nice place.)

A 'point' so unimaginably small there isn't even a mathematics equation that can define it as existing. A 'point' of zero volume. There is no 'space' (or space-time) left as we know it. Yet this point of zero volume holds nearly all of the mass that existed in the core of our original star. And perhaps more. Because as the singularity forms, its colossal gravitational power sucks in other surrounding matter, such as gas clouds, comets, whole stars and, as if that wasn't enough, this monstrous sucking abomination even consumes other smaller black holes, if they are close. A cosmic Jabba the Hutt, feasting and gorging on anything foolish enough to enter its feeding ground.

The simplest analogy to give you a mental picture of these cosmic monsters is a sink and plughole.

Picture a sink filled with water. Pull out the plug and the water will begin to drain down the plughole, spinning around and around as it goes, speeding up as it gets nearer the hole. The water in this analogy is like space-time and matter combined. The increasing density of the matter causes space-time to wrap itself in and around it, causing both to spiral downwards faster and faster, into an ever-smaller space. Where it differs from the sink analogy is that the imploding matter and space-time 'create' the hole. There isn't a cosmic plughole waiting for matter and space-time to fall into it. The imploding matter and warped and twisted space-time are what make the 'hole'!

And, while a plughole directs water neatly into your drains, the hole created by matter and space-time just keeps going down, as the condensed matter and intertwined space-time try to squeeze into an ever more small and confined space until they reach a point where the curvature of space-time becomes infinite. (And infinite is a lot. More than any lot you can think of.) And the density becomes infinite. (There is no analogy to describe this kind of density.) As a consequence, the power of gravity becomes infinite. And every law of physics that we know of breaks down, or – if there is any sense in the phrase – 'infinitely breaks down'. While we can give this nightmarish phenomenon a name, a 'singularity', we can never see it. Ever.

Because in any region of space where a singularity is forming, light (in the form of light photons) is forced to follow the curvature of space-time. (Remember light travels 'through' space, so if space is curved, light has no choice but to follow that curvature.)

Go back to our sink analogy for a minute to understand what happens to light. Imagine on the edge of our sink is a little ant. He's standing on the sink's edge, just a fraction of a millimetre away from the water, as the plug is pulled. While we're all shouting, 'Don't do it!' he takes an ant step forward and plunges into the swirling water as it spirals down, faster and faster, towards the waiting plughole. No amount of effort or energy that little ant could ever exert would be enough to save him from the whirlpool of water. It carries him down and down until the poor little fella eventually disappears into the 'plughole abyss' never to be seen again. A forming singularity does to light what the 'sink-water-plughole' does to our ant.

As a light beam enters a region of space-time where a singularity is forming, the monstrous gravitational power of the singularity is spreading outwards in all directions into the surrounding space. This creates a violent vortex of twisting, churning space (like the swirling water in our sink) that is spiralling downwards, faster and faster, towards the singularity. The edge of this violent vortex has a name, it's called the 'event horizon'. The event horizon marks an area in space-time where light can either resist the black tentacles of the singularity (and so fly on by) or get sucked in and down. If light can't resist the gravitational pull of the singularity and is forced across the event horizon boundary, then it's just like our ant stepping forward off the edge of the sink. It's a one-way ticket down into cosmic hell.

The 'only' difference between our ant and light is we can see our little ant spiralling down to be consumed by the sink's 'hole'. But when light does its 'stepping-off-the-sink-edge' impression and crosses the event horizon, it disappears instantly. Goes out. Doesn't shine. Is 'kaput'. Because inside the boundary around a singularity, inside the event horizon, space-time is twisted and curved to such an extent and fused with such huge concentrations of condensed, imploding matter, that light is sucked down faster than light itself can travel. As much as it might try, light can't fight its way back up. Like a drowning man in concrete boots, watching the surface of the water slowly disappearing as he sinks down and down into the murky depths, so any light crossing the event horizon is instantly sucked down and extinguished by the gravitational vortex that lies inside.

Can you possibly imagine a force of gravity that can do this – trap light?

Remember in Chapter 5, Space and Time, we described how space-time curves in and downwards towards Earth, because of the presence of Earth. To escape Earth's surface, a rocket has to travel at a speed of seven miles a second, or 25,000 mph (40,000 kph) to break free of this 'curved downwards' space. This is called 'Earth's escape velocity'. Now think about the gravitational grip a singularity exerts on the space-time around it, all those googolzillions of matter playing cosmic sardines in a 'space' too small to even have a measurement. It's so immense that light, travelling at 671 million mph, hasn't got enough speed to break free once inside the event horizon. The escape velocity inside the event horizon is too much for light. The fastest known thing in the whole of the universe is reduced to a damp squib.

As a consequence, everything inside the event horizon is... black. A black hole in space that we can never see into. What's in there? We don't know. It's not matter as we know it. Every particle that we understand as making up the living world (read 'universe') has gone.

It's not space as we know it. Space has a volume, up, down, across. According to the best brains we have, the theoretical singularity at the centre of a black hole has no recognisable volume. It's not time as we know it. The warping of space-time at the event horizon, the 'edge-of-the-hole', is so extreme, time comes to a virtual stop. So, inside the event horizon, there is unlikely to be any time as we understand it. No ticking. And if all that wasn't nightmare-inducing enough, 'the hole' doesn't communicate. Other objects, like stars, comets, moons or gas clouds, tell us stuff through the light they express or their movements or sound. The inside of the hole, the inside of the event horizon, gives out no information.

There is no light.

No movement to see.

No sound.

It tells us nothing about what's going on 'inside'.

Black quicksand.

If we were to take a guess at what all the quintillions and septillions of tonnes of matter that get sucked in turn into, it's likely to be some kind of primordial energy (remembering Einstein's equation, $E = mc^2$, which tells us mass (matter) and energy are one and the same thing, just in different forms). In the neutron star's centre we do have an idea of what pulverised matter ends up like – the creeping crawling gets-out-of-an-airtight-container-on-its-own (it's behind you!) superfluid. But in a black hole and at the singularity we have no real idea what kind of energy matter and, indeed space-time, transitions into when it's been flayed of all its form and substance. Suffice to say, it's going to make the neutron star's superfluid look like a tub of organic yoghurt.

But if we know nothing of a black hole's 'insides', we do know something of what we see of them on their outside. And that is, they come in sizes of big. Seriously, cheek-clenchingly big. And there's a seriously big one right at the centre of our galaxy, the Milky Way. It's called Sagittarius A. Remember how big our Sun is? You could fit 1.3 million Earths inside it (so huge!). Well, the black hole at the centre of our galaxy is four million times more massive than our Sun! (A black hole four million times bigger than our Sun!! In <u>our</u> galaxy!)

Its diameter (the measurement across its event horizon) is around 27 million miles (44 million km). That's not too far off the distance from Earth to Mars! And it's all 'hole' (slight trembling of the knees at this point). A 27-million-mile-across black hole. And we know nothing, absolutely nothing for certain, about what (or who?) is inside. Be very thankful for light years. There's 26,000 of them between us and 'it'. But we said sizes of big. And the hole at the centre of our galaxy, whilst it's called a 'supermassive black hole' (I don't really want to hear the word 'supermassive'), doesn't do big, big. That lurks in the pitch black elsewhere (please be far away!). This huge hole is called S5 0014+81 (or 'huge hole' for short). The figures for this hole are deeply disturbing. In fact, I'm not sure I'll ever sleep again.

Our 'supermassive black hole', at the centre of our galaxy, has a mass equivalent of four million Suns. S5 0014+81 has a mass equivalent of... 40 BILLION Suns! (Plus another 39 exclamation marks!)

And if that wasn't big-time scary enough, its horrific gravitational power reaches out, sucks in and devours the equivalent of 4,000 times the mass of our Sun each year, in the form of gas clouds, planets, stars and other black holes. And I'm just going to put its diameter out there, for you to convulse over... it's 146,000,000,000 miles WIDE! Yes, that's BILLIONS! (Or 236,000,000,000 billion km.) How big is that? Do you really want to know?

In Chapter 2 'How Far is Far?' we went to Pluto, close to the edge of our solar system. Our solar system is 7.5 billion miles in diameter (12 billion km). Well, big, increasingly scary 'huge hole' is 20 times BIGGER than our whole solar system. And it's a deep, dark, never ending HOLE, in space! One hundred and forty-six billion miles across. It could suck our whole solar system, plus 19 others of a similar size, into oblivion in just one huge gulp (gulp!). Stand in front of a mirror, open your mouth as wide as you can and hold up a pea in front of it. Your mouth is S5 0014+81. The pea is our whole humongously big solar system (and in that pea, somewhere, is a billion-times-less-than-a-barely-visible-pinprick – you).

The good news, thankfully – oh, so very thankfully – is 'huge hole' is 12.1 billion light years away in another galaxy (hurrah!!). The bad news is, because it's taken 12.1 billion years for light from S5 0014+81 to reach us, we are seeing it as it was just 1.7 billion years after the Big Bang. It's had another 12 billion odd years to grow... even BIGGER. (Four thousand of our Suns consumed each year, times 12 billion years, equals... oh, dear.)

Now, it's worth making a note here, were you ever to get into a pub or dinner party conversation about black holes and singularities. (I mean, come on, what else is there to talk about?) We mentioned in Chapter 5 'Space and Time', the emerging theory of loop quantum gravity (LQG). Well, this theory questions the existence of the singularity at the centre of black holes as being infinite, both in terms of the curvature of space-time and density. Because as we know, LQG theory predicts space-time itself is granular. That is, just like we understand matter is made up of atoms, and those atoms drill down to eventually land at ever tinier things (until you reach the quarks) so space-time 'drills down' to eventually land at infinitesimally tiny 'bits' that make up space-time.

Carlo Rovelli, in his brilliant book, *Reality Is Not What It Seems*, describes these tiny bits as 'nodes'. These 'nodes' are individual volumes of space, connected to each other by equally small things called 'links', that together create tiny 'loops'. And these loops, as we said earlier, strung together in their endless multitudes, give us what we experience as three-dimensional space. We should remember here these loops are not 'in space', they are what space itself is made of, at its most granular level. Their size is Planck length size. That's this size: 0.000 00000000000000000000000000000001m. (That's 34 noughts, to save you counting. Oh, and don't forget the '1' at the end!)

How small is this?

Well, take a marble and blow it up until it's the size of the visible universe. That's a marble 93.4 billion light years in diameter. Even at this scale of magnification, the Planck length is so small it would still not be visible (which is truly astonishing). It would still be a million times smaller than the marble was, before magnification. That's how small the Planck length is (that's a Carlo Rovelli analogy, and a good one). This Planck length, the predicted size of nodes and links that make up loops, is an indivisible size. The ultimate, finite small. Nothing can or could exist beyond this scale.

How is this relevant to black holes and singularities? Well, LQG theory says that singularities are not infinite in density and don't cause space-time to curve infinitely. Because the gravitational collapse that went big star – white dwarf – neutron star – black hole –singularity, has to stop somewhere! It can't keep collapsing downwards indefinitely. And, as far as LQG theory is concerned, the collapse, with all its increasing density and curving space-time, finally stops as space itself reaches its smallest point – the loop, the Planck length scale loop. The size of the individual grain that space-time itself is constructed from.

Now, down at this size you may be wondering whether something as ridiculously small as the Planck length versus a never-ending 'infinitely small' has any meaning in the real world? Well, yes, it does. Because, if all the collapsing matter of a giant star and all the other planets and stars a black hole may have gobbled up eventually reaches a point

(Planck length scale) where it can't collapse any further (i.e. infinitely), then theory says it will reach a point of pressure where it bounces, or rebounds. And all the humongous, compressed, dense 'fluid-stuff' that matter has been ground into over billions of years will come flying back out! So, worth a small diversion, just so you can be on the lookout for that particular occurrence, should it happen.

Finally, if we can't see them, how do we know black holes are there? Two main ways. One not so dramatic, the other absolutely dramatic.

The first is the way we can see gas clouds, planets and stars being dragged in the direction of an area of space where, to our eyes (that will be telescopes), there is nothing there to do the pulling. It's just a region of dense blackness. Us laymen now know, of course, that will be a black hole exerting its immense gravitational influence on surrounding stars and the like. The second way is by something called an 'accretion disc'. Just as the water in our sink analogy spins faster and faster as it reaches the plughole, so does all the debris from stars and gas clouds as they fall towards the black hole and are systematically ripped to shreds. These vast quantities of debris (matter) begin to spin around the black hole's outer edge, its event horizon, getting faster and faster. In some instances, this whirlpool of frenzied spinning debris can reach speeds of billions, even trillions, of miles per hour. The combination of this increasing speed, and the matter also becoming increasingly compressed, creates friction, which releases gargantuan bursts of energy in the form of X-rays and gamma rays.

Until the mad spinning debris crosses the event horizon (never to be seen spinning again), these X-rays and gamma rays radiate outwards into the depths of space. Such is the energy release of a gamma ray that the resultant light can be as bright as all the stars in a hundred normal size galaxies. That's the brightness of trillions upon trillions of stars. This burst of light from a gamma ray is one of the most distant things we can see in the universe, because of its brightness. (The name given to this burst of energy is a 'quasar'.)

It's a strange thing. One of the brightest things to ever happen in the universe, a burst of light brighter than all the stars in a hundred galaxies combined, happens just before all the mad spinning material causing the heat and energy release, crosses the black hole's threshold of no return, the event horizon. And all the lights go out. It's as if all the things we know of that make up this world – matter, energy and light – crunch together to scream a final 'Hurrahhhh!' before falling into the abyss. (Or it is a warning?)

A last thought.

You will remember in Chapter 5 we talked about the way space curves in the presence of matter. And how curved space around Earth lessens in intensity the further away it is from Earth, but that the curving effect never goes to zero. No matter how far away any part of the universe might be from Earth, the space in that part of the universe still 'feels' the impact of Earth's presence.

Well, it's the same with black holes.

As well as Sagittarius A, the supermassive, 27 million miles wide black hole at the centre of our galaxy, there are estimated to be close on another 100 million smaller black holes in our galaxy alone. (Yes, 100 MILLION!) Then, of course, there's the nightmare hole, S5 0014+81, all 146 billion miles of it, 20 times bigger than our solar system. The monstrous gravity of S5 'huge hole' and all the millions of black holes in our galaxy will all be tugging on you.

Tonight.

As you go to bed.

The bogeyman is truly in the room.

Sleep tight.

CHAPTER 8

WHO ARE WE?

So there you have it. We're on an infinitesimal 'ot' lost in an infinity of blackness (we don't even warrant the word 'dot'). We've been on this tiny rock as reasonably intelligent, conscious beings for roughly two million years. And yet we still don't know how, or where, the universe that spawned us came from. We don't know where it's going; we can't, and never will be able to see the edge of it, if indeed it has an edge. And we don't know what makes up the vast majority of the universe we exist in.

Something called 'dark matter' makes up around 27 per cent of all matter in the universe. We can't see it directly. We don't know what it is. We 'think' it's there. Something called 'dark energy' makes up 68 per cent of the universe. We're pretty sure it's the thing that's making the universe not just expand, but expand at an ever faster and faster speed. But again, we can't see it, we don't know what it is. The stuff that we can see and do understand accounts for just 5 per cent of the total energy and matter in the universe (so, what we think of as normal, is actually nothing like the norm!). The other 95 per cent is a complete mystery (stuff for another layman's journey, perhaps?). The biggest mystery of all, however, is we don't know who we are or why we are here.

A few random 'boldly' thoughts to finish on, in response to the last question.

1. We're a prison. The rest of the universe is populated with super-intelligent beings who have learnt to be nice to each other. The few amongst them who transgress from 'niceness'... they send them here, to the place called 'Earth'. The reason we can't see, find or make any contact with any of these other forms of life is these super-intelligent, reasonably nice beings put Earth as far out into infinite space as possible to make our existence as cold and isolated as possible. To ensure there is no possibility we can contaminate any other cosmic civilisation.

Mad?

Well, think about it. We emerge on Earth in an infantile state, screaming our heads off. The person who has to bear us screams their head off too. We grow to kill, maim, torture and abuse, generally living life snarling at each other, getting more and more bitter and disgruntled as we age. And, if we get beyond a certain quota, Earth erupts, explodes, cracks and floods with a degree of regularity that keeps our numbers in check. Earth as a cosmic version of Alcatraz might not be such a daft idea.

2. We're a cosmic version of the National Lottery. We're a one-off, on our own. We beat the odds of life existing in the universe, which has been calculated at around one in ten billion, trillion. And, because lightning doesn't strike twice (unless you're Mike McDermott from Portsmouth, who has won the UK lottery twice with the same numbers! The odds on the first win were two million, three hundred and thirty thousand to one. The odds on the second win, with the same numbers, were a staggering five trillion to one! And yet, he did it), we're not going to meet any other folk. The universe conjured up a never-to-be-repeated 'miracle'. It created a cocktail of chemicals, gave it a stir for around 13 billion years and suddenly 'it could be you' became us.

3. We're just an average planet, with averagely matured life, amongst trillions of other equally average planets and life forms. After 13 billion years, life is just a natural part of cosmic evolution. It happened everywhere at roughly the same time. And it continues to happen. In terms of technological progress, however, we're still at the 'Neanderthal' stage. After two million years of existence, we've still only got to our Moon, a trifling distance of just 225,000 miles.

Just getting to the edge of our tiny solar system (over three billion miles) with a manned space flight looks hundreds, if not thousands, of years away. To get to the next nearest planet outside of our solar system (25 trillion miles) we're talking around 150,000 years of travel time at our current speeds.

And it's the same for all the 'other' people, aliens, 'things', on all the other planets. The damn universe is just too big. Worse, it's growing at a rate close to or faster than the speed of light. None of us are going to get to meet up for aeons (however long that is).

4. We're not alone and we're not average. Unfortunately for us, we're way below average. We started late, which puts us on the bottom rung of the cosmic evolutionary ladder. There are 'super-beings' out there. Thousands, possibly hundreds of thousands of years more advanced than us. They've gone to their moon. To the edge of their solar system. They've crossed their galaxy and even made it to the next galaxy.

They're now eyeing up the other two trillion odd galaxies around them to see which ones they fancy paying a visit to. Only the ludicrous distances involved in travelling the universe (and the fact the damn thing won't stop expanding!) have kept them from stumbling over us. But they're on their way. (The late and brilliant Stephen Hawking thought this a reasonable, if not probable, scenario... have I gulped before?)

5. We're a simulation. The universe, space-time, all that's in it, including Earth and 'us', are no more than a computer simulation. This would mean that time, which we think came into existence with the Big Bang, 13.8 billion years ago, didn't. Time is actually infinite. And, at some time in

time's infinite past, a very super-intelligent alien race of some kind created the universe as a mathematical-based computer simulation.

After running it for a while and perhaps getting bored with only space, energy, stars and black holes to play with, this super-intelligent alien race changed the code and created us as an additional plaything (a new product feature – create two, get seven billion FREE!).

Bonkers?

Well, a pretty well-respected physicist called Max Tegmark has written books and talked a great deal on this subject, and he says that the closer he gets to analysing the smallest fundamental particles that make up everything in the universe, things like quarks and electrons, the more it appears they move around and behave to rules that are entirely 'mathematical'. The same mathematical rules which we use to create game simulations in 'our world'.

So, the next time someone tells you to 'get real', take a closer look at them. They might not be who they appear to be.

6. And, finally. You are no more than my imagination. I'm a single, super, super, intelligent entity who did in fact 'create the universe'. Just like you need some playthings to keep you amused – dolls, teddy bears, drink, gambling, Elf on the Shelf – so did I. So I invented you for my amusement. You're just all made up characters in my super-being head.

I am, after all, 'boldly layman'.

Now that one certainly feels the most plausible to me.

LAYMAN'S LOG

You look up at the night sky and marvel at the twinkling multitude.

But now, as you know, even the nearest star beyond our Sun would take 150,000 human years to travel to. That's a timespan so great that to be arriving there now, you would have to have set off when it's estimated there were fewer than 2,000 of our species of primate on Earth, struggling to avoid extinction in sub-Saharan Africa.

But, just as the impossibility of star travel creeps into your soul, a journey that won't be a reality for thousands of generations, one morning someone says, 'It's behind you.' And you turn to see the most monstrous of stars rising above the horizon, flooding your world with light. A star so large it could gobble up one million, three hundred thousand Earths. A star that's been alive for 4.6 billion years and burns, at its core, to temperatures in excess of 15 million degrees Celsius.

Our star. The Sun.

We are one of those flickering, magical specks of light floating against a backdrop of infinite darkness. To any other distant eyes, we are the planet Proxima b circling the star Proxima Centauri.

What you look at, and marvel at, is here.

REFERENCES

Endnotes

[1] *The New York Times*, opinion letter to the editor from Dorothy C. Morrell, Seattle, 18 September 1986, https://www.nytimes.com/1986/09/28/opinion/l-just-how-long-is-a-trillion-seconds-229186.html.

[2] Gary Ernest Davis, *Republic of Mathematics*, [web blog], 16 February 2010, blog.republicofmath.com.

[3] Tim Urban, 'From 1,000,000 to Graham's Number', *Wait But Why*, [website], 20 November 2014, https://waitbutwhy.com/2014/11/1000000-grahams-number.html.

[4] Table of Large numbers-Sunshine.chpc.utah.edu>ManSciNot1.

[5] T.J. Cox and Abraham Loeb, 'The Collision Between the Milky Way and Andromeda', Harvard-Smithsonian Center for Astrophysics, originally at arxiv.org.

[6] K. P. Schroeder and R. C. Smith, 'Distant Future of the Sun and Earth Revisited', *Monthly Notices of the Royal Astronomical Society*, 2008, http://sro.sussex.ac.uk/1800/2/DoomsdayResub.pdf.

[7] D. Carrington, 'Date Set for Desert Earth', *BBC News Online*, [website], February 2000, http://news.bbc.co.uk/1/hi/sci/tech/specials/washington_2000/649913.stm.

[8] Bradford G. Schleifer, 'Designed for Discovery, Part 3: The Universe', *The Real Truth*, https://rcg.org/realtruth/articles/070803-003.html.

[9] Doctor Richard Egglestone. Amazing equations to work through the Ant/Atlantic Ocean analogy.

[10] John D. Norton, 'How big is an atom?', *University of Pittsburgh*, [website], 17 June 2006, https://www.pitt.edu/~jdnorton/Goodies/size_atoms/index.html.

[11] Doctor Richard Egglestone kindly worked out 'The Millimetre Challenge'.

[12] 'Quantum Foam', *NASA Science Beta*, [website], 31 December 2015,https://science.nasa.gov/science-news/science-at-nasa/2015/31dec_quantumfoam.

[13] 'Time Dilation at "Low" Speeds', *E = mc2 Explained*, [website], http://www.emc2-explained.info/Time-Dilation-at-Low-Speeds/#.WumKA4gvzIU.

[14] John Walker, 'C-ship: The Dilation of Time', *Fourmilab*, [website], https://www.fourmilab.ch/cship/timedial.html.

[15] John Walker, 'C-ship: The Dilation of Time', *Fourmilab*, [website], https://www.fourmilab.ch/cship/timedial.html.

[16] 'Sun Facts', *Space Facts*, [website], https://space-facts.com/the-sun/.

Printed in Poland
by Amazon Fulfillment
Poland Sp. z o.o., Wrocław